SpringerBriefs in Applied Sciences and Technology

Nanotheranostics

Series Editors

Subramanian Tamil Selvan, Institute of Materials Research and Engineering, National University of Singapore, Singapore, Singapore

Karthikeyan Narayanan, Singapore, Singapore

Padmanabhan Parasuraman, Singapore, Singapore

Paulmurugan Ramasamy, Stanford University School of Medicine, Palo Alto, CA, USA

Indexed by SCOPUS. Nanotheranostics is a burgeoning field in recent years, which makes use of "nanotechnology" for diagnostics and therapy of different diseases. The recent advancement in the area of nanotechnology has enabled a new generation of different types of nanomaterials composed of either inorganic or polymer based nanoparticles to be useful for nanotheranostics applications. Some of the salient features of the nanotechnology towards medicine are cost reduction, reliable detection and diagnosis of diseases at an early stage for optimal treatment. The advent of nanotheranostics is expected to benefit the pharmaceutical and healthcare industries in the next 5-10 years. Nanotechnology holds a great potential to be explored as a multifunctional platform for a wide range of biological and engineering applications such as molecular sensors for disease diagnosis, therapeutic agents for the treatment of diseases, and a vehicle for delivering therapeutics and imaging agents for theranostic applications in cells and living animals.

Subramanian Tamil Selvan

Nanomedicine

Emerging Prospects

 Springer

Subramanian Tamil Selvan
Alpha Biomedical Pte. Ltd.
Singapore, Singapore

ISSN 2191-530X   ISSN 2191-5318 (electronic)
SpringerBriefs in Applied Sciences and Technology
ISSN 2197-6740   ISSN 2197-6759 (electronic)
Nanotheranostics
ISBN 978-981-99-2138-6   ISBN 978-981-99-2139-3 (eBook)
https://doi.org/10.1007/978-981-99-2139-3

This Springer imprint is published by the registered company Springer Nature Singapore Pte Ltd.
The registered company address is: 152 Beach Road, #21-01/04 Gateway East, Singapore 189721,
Singapore

Contents

Chapter 1
Introduction to Nanomedicine

Abstract A brief introduction is given to nanomedicine, an emerging paradigm intersecting two burgeoning fields of nanotechnology and medicine. It covers the application of nanomedicine in diagnostics and therapy of a wide range of diseases such as cancer, cardiovascular, orthopaedics, and neurodegenerative disorders. A wide range of nanomaterials, nanoparticles and biomaterials for these applications are discussed.

Keywords Nanoparticles · Nanomedicine · 2D nanomaterials · Quantum dots · Diagnostics · Therapy · Cancer · Neurodegenerative disorders · Cardiovascular diseases · Orthopaedics

1.1 Nanomedicine

Nanomedicine is a field of interdisciplinary science that integrates physical, chemical, and engineering sciences, utilizing nanotechnology (functional nanomaterials, and structures at the nanometer scale between 1 and 100 nm) and medicine (drugs, imaging tools and delivery devices) for disease diagnosis and therapy.

Today, nanomedicine is a buzzword for a variety of diseases including cancer (Chow and Ho 2013; Min et al. 2015; Chen et al. 2017; Liu et al. 2017; Nam et al. 2019), cardiovascular (PA Ferreira et al. 2015; Di Mauro et al. 2016), orthopaedics (Mazaheri et al. 2015; Perli et al. 2017), dental (Besinis et al. 2015; Padovani et al. 2015; Chieruzzi et al. 2016; Priyadarshini et al. 2016; Fawzy et al. 2017; Priyadarshini et al. 2017), kidney (Marom et al. 2012; Kamaly et al. 2016; Williams et al. 2016), and neurodegenerative diseases (Goldsmith et al. 2014; Saraiva et al. 2016; Tapeinos et al. 2017; Teleanu et al. 2019a).

S. Tamil Selvan, *Nanomedicine*, Nanotheranostics,
https://doi.org/10.1007/978-981-99-2139-3_1

1.1.1   Nanomaterials for Cancer Nanomedicine

In cancer nanomedicine, a wide range of nanomaterials including two-dimensional 2D MoS_2/Bi_2S_3 (Liu et al. 2014; Wang et al. 2015a, b; Song et al. 2016), MnO_2 nanosheets (Chen et al. 2014f), graphene oxide (Chen et al. 2014e), transition metal dichalcogenide nanomaterials (Gong et al. 2017) have been developed extensively for therapeutic and diagnostic (i.e. theranostics) applications of cancer (Peng et al. 2017). Nanotechnology assisted approaches for stem cell differentiation, tracking, labelling, and therapy have been delineated in recent reviews by our group (Nanda et al. 2017; Yi et al. 2017).

Different nanoparticles (NPs) have been designed for nanomedicine over the last decade. Metallic NPs (e.g. Au, Ag, Pd, Pt, Cu) have been used as plasmonic nanosensors or surface-enhanced Raman scattering (SERS) probes for label-free ultrasensitive molecular detection of body fluids (Kosaka et al. 2014; Bui et al. 2015; Lane et al. 2015; Langer et al. 2015; Jeong et al. 2016; Yang et al. 2016b; Xie et al. 2017). Conversely, semiconductor quantum dots (QDs) have been extensively used for biological applications (Mattoussi et al. 2000; Gao et al. 2004; Medintz et al. 2005; Chang and Rosenthal 2012).

Despite toxicity issues related to heavy metal cadmium, even today, semiconducting NPs such as CdSe/ZnS QDs are the best diagnostic agents for in-vitro cell labelling and in-vivo animal imaging studies, thanks to their excellent optical properties and stabilities (Yen and Selvan 2015; Freyer et al. 2019; Hanifi et al. 2019; Ondry et al. 2019) Alternate non-cadmium based QDs have emerged in response to combat heavy metal Cd-based cytotoxicity (Xu et al. 2016). For example, Mn-doped ZnS QDs have been used as protein sensors (Wu et al. 2013), used for detection of H_2S (Wu et al. 2014) and dopamine (Diaz-Diestra et al. 2017) in biological samples, and as imaging probes for intracellular Zn^{2+} ions (Ren et al. 2011). Earlier, our group contributed to the grafting of Mn-doped ZnS nanocrystals and anticancer drug (doxorubicin) onto graphene oxide for cell labelling and delivery (Dinda et al. 2016). Conversely, molybdenum disulfide QDs (Liu et al. 2018) has been used for the detection of dopamine.

Recently, ZnO nanowires and nanocomposites (e.g., Ag–ZnO) have shown great potentials in the detection of cancer biomarkers such as RNA, DNA, proteins, and extracellular vesicles (Guo et al. 2018; Paisrisarn et al. 2022; Chattrairat et al. 2023; Huang et al. 2023; Jung et al. 2023). It is worth mentioning here the application of ZnO and TiO_2 nanostructures for the biosensing of proteins using the surface-enhanced Raman scattering (SERS) approach (Adesoye and Dellinger 2022).

Multifunctional NPs for multimodal bioimaging incorporating optical imaging using NIR emitting QDs or up-conversion luminescence, computed tomography (CT) and magnetic resonance imaging (MRI), and therapy (Lee et al. 2012) (photodynamic, photothermal, targeted drug delivery (Liu et al. 2015), pH-triggered on-demand drug release (Wang et al. 2015c) etc.), have attracted immense interest (Chen et al. 2014d; Wu et al. 2015; Li and Chen 2016; Duan et al. 2017; Amirav et al.

2019). We have also pioneered the synthesis of bifunctional nanomaterials (fluorescent QDs, magnetic iron oxide, up-conversion, and magnetic/antibacterial NPs) for bimodal imaging (optical and MRI) and therapeutic applications (Selvan et al. 2007; Ang et al. 2009; Selvan et al. 2009; Das et al. 2010; Selvan 2010; Zhang et al. 2014). Carbon nanodots (Bhunia et al. 2013; Shi et al. 2015; Xu et al. 2015) and graphene QDs (Zhang et al. 2012; Zheng et al. 2015a, b; Yang et al. 2016a; Yao et al. 2016; Yan et al. 2019) have been extensively explored as bioimaging probes. Interestingly, carbon dots have recently emerged as a potential candidate system in nanomedicine to protect the cells from oxidative stress, eliminating intracellular reactive oxygen species (ROS) (Xu et al. 2015). It is worth mentioning here the use of ceria–zirconia NPs as a therapeutic nanomedicine for treating ROS-related inflammatory diseases such as sepsis (Soh et al. 2017). Several ROS-mediated nanomedicine systems have been delineated recently (Yang et al. 2019; Ding et al. 2023; Naik and David 2023).

Notable advances have been made in the synthesis of different magnetic NPs (MNPs) (e.g., Fe_3O_4, Fe_2O_3, FePt, Co), and their nanostructures and composites. (Yen et al. 2013b; Yen et al. 2015; Kang et al. 2017; Wang et al. 2018; Yang et al. 2018; Ray et al. 2019; Satpathy et al. 2019; Esthar et al. 2023; Liu et al. 2023).

These magnetic nanocomposites can be used as drug carriers (Farmanbar et al. 2022; Turrina et al. 2022; Esthar et al. 2023; Liu et al. 2023), hyperthermia agents (Ansari et al. 2022; Shabalkin et al. 2023), and MRI contrast agents (Cheraghali et al. 2023; Jiang et al. 2023) in cancer diagnosis/bioimaging (Mohapatra et al. 2023) and therapy (Su et al. 2023; Vangijzegem et al. 2023). Iron oxide NPs combined radioisotopes (e.g., Tc-99 m) can be used as dual modality contrast agents for the high spatial resolution of MRI applications combined with high sensitivity single photon emission computed tomography (SPECT), and positron emission tomography (PET) (Karageorgou et al. 2023).

Upconversion NPs (UCNPs) (e.g. $NaYF_4$:Er, $NaGdF_4$:Er) are another interesting class of materials utilized extensively for bioimaging, owing to their stable luminescence; and fabricated as core–shell NPs (Dou et al. 2015) or multifunctional NPs for bioimaging, and photodynamic therapy (PDT) (Idris et al. 2012; Chen et al. 2014a; Wang et al. 2015b; Zhou et al. 2015; Zhou et al. 2016; Xu et al. 2017; Liu et al. 2019b; Zhang et al. 2019). Other polymeric NPs (Ang et al. 2014), hybrid NPs (Nguyen and Zhao 2015; Zhang et al. 2017), and multifunctional NPs derived from small organic building blocks (Xing and Zhao 2016) have considerably contributed to nanomedicine. Conversely, rare-earth oxide NPs (e.g. gadolinium oxide) found their potential uses in MRI and chemotherapy (Wu et al. 2019).

Although metallic NPs such as Au, Ag are synthesized in water directly, most of other NPs such QDs, UCNPs, MNPs are synthesized in presence of organic ligands at temperatures over 200 °C, resulting in hydrophobic NPs. Different coating methods have been developed to make these hydrophobic NPs water soluble. Today, the stabilization of NPs in water and biological media has become a matured strategy, thanks to a wide variety of coating strategies that exist in the literature. This includes silica (Mulvaney et al. 2000; Gerion et al. 2001; Selvan et al. 2004; Darbandi et al. 2005; Selvan et al. 2005; Yi et al. 2005; Zhelev et al. 2006; Tan et al. 2007), polymer (Hong et al. 2012; Wang et al. 2013; Yen et al. 2013a; Chen et al. 2014b; Topete et al. 2014;

Palui et al. 2015), peptides (Narayanan et al. 2013; Chen, Li et al. 2014c; Yang et al. 2017; Zhang et al. 2018), lipids/liposomes (Medintz et al. 2005; Murcia et al. 2008; Weng et al. 2008; Al-Jamal et al. 2009; Tian et al. 2011), proteins (Mattoussi et al. 2000; Chithrani and Chan 2007; Yang et al. 2013; Hu et al. 2014; Tay et al. 2014; Sasaki et al. 2015; Scaletti et al. 2018), antibodies (Goldman et al. 2002; Medintz et al. 2005; Snyder et al. 2009), and enzymes (Kong et al. 2016) for the stabilization of NPs. Hydrophobic ligands such as HDA can also be used for the stabilization of Au NRs and heterostructures (Cheng et al. 2014; He et al. 2014).

1.2 Challenges and Advancements of Nanomaterials for Nanomedicine

In general, nanomaterials in biomedical applications pose an important concern: what are the safety concerns of nanomaterials? How do we address the growing needs of ageing population with neurodegenerative disorders, and early diagnosis and therapeutic measures for diseases like cancer? This Book attempts to address the above concerns with the advent of nanomedicine. Compared to cancer nanomedicine, the application of nanomedicine in neurodegenerative diseases is still in its infancy state. The biggest challenge in neurodegenerative diseases is to tackle the permeability of blood-brain-barrier (BBB) and deliver therapeutic drugs to the brain (Ramanathan et al. 2018). Toward this goal, nanoscale materials have been developed and used either as bio-labelling agents or as therapeutic carriers, and in some cases as neuro-protective agents for neurodegenerative diseases (Goldsmith et al. 2014; Saraiva et al. 2016; Teleanu et al. 2019b; Liu et al. 2019a; Le Floc'h et al. 2019).

This Brief focuses mainly on the application of nanomedicine in cancer and neurodegenerative diseases. It also attempts to cover the application of nanomedicine in other emerging areas such as orthopaedics, and cardiac diseases (Fig. 1.1).

1.2.1 Nanomedicine Advancements in Cancer and Neurodegenerative Diseases

Some of the recent advancements (See Chap. 2) in cancer diagnosis (e.g., multimodal tumor imaging) and therapy (e.g., combined therapies involving either photothermal, chemotherapy, photodynamic or immunotherapy) have been made using 2D nano-materials (Chen et al. 2020; Ding et al. 2020) (e.g., MoS_2/Bi_2S_3 nanocomposites (Wang et al. 2015a; Wang et al. 2019), molybdenum oxide nanosheets (Song et al. 2016; Wang et al. 2023b), MoS_2 nanosheets (Liu et al. 2014; Murugan and Park 2023), MnO_2 nanomaterials (Chen et al. 2014f; Tan et al. 2017; Ding et al. 2020), doped graphene nanosheets (Lu et al. 2022), graphene oxide-based multifunctional nanomaterials (Gonçalves et al. 2013; Chen, Xu et al. 2014e; Gu et al. 2019; Itoo

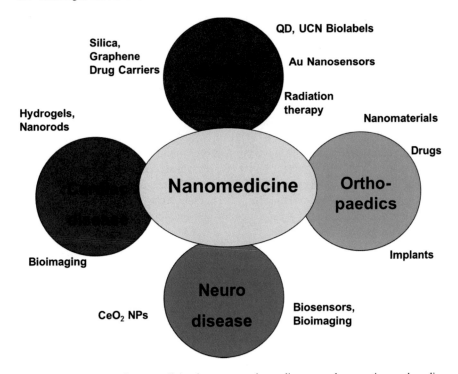

Fig. 1.1 Applications of nanomedicine in cancer, orthopaedics, neurodegenerative, and cardiac diseases

et al. 2022), and multifunctional Au-based nanomaterials (Ouyang et al. 2023; Wang et al. 2023c), and magnetic nanomaterials (Mukherjee et al. 2020; Liu et al. 2021).

Chapter 3 deals with different nanomedicine approaches for neurodegenerative diseases such as Alzheimer's disease (AD). Design of inorganic NPs (e.g., Au, ZnO, MoS_2, CeO_2) and organic NPs (e.g., curcumin, green tea polyphenol- EGCG) for inhibiting the amyloid aggregation and tau hyperphosphorylation associated with the AD are discussed (Han et al. 2017; Shukla et al. 2021; Tamil Selvan et al. 2021). Different NP-based drug delivery approaches (e.g., apolipoprotein, peptides, dendrimers) to the delivery of CNS drugs across the blood–brain barrier (BBB) are also discussed (Tapeinos et al. 2017; Arvanitis et al. 2020; Loch et al. 2023).

1.2.2 Nanomedicine Advancements in Orthopaedics and Cardiovascular Diseases

Chapter 4 addresses nanomedicine and tissue engineering approaches for orthopaedics. Bone mimicking scaffolds composed of polymers (e.g., polycaprolactone, polylactic acid, chitosan) and inorganic nanomaterials (e.g., reduced graphene oxide rGO, hydroxyapatite) (Seyedsalehi et al. 2020), and zwitterionic chitosan/β-tricalcium phosphate hydrogel/GO scaffolds (Wang et al. 2023a) for bone tissue engineering applications are covered. Orthopaedic drug delivery systems using dextran/β-tricalcium phosphate nanocomposite hydrogel scaffolds (Ghaffari et al. 2020), and chitosan-vancomycin hydrogel bone repair scaffold (Gao et al. 2023) are also delineated.

Chapter 5 delineates the applications of nanomedicine in diagnostics and treatment of cardiovascular diseases (CVDs). Recent developments in multifunctional NPs (Kleinstreuer et al. 2018), nano/biomaterials and devices to diagnose and treat a variety of CVDs with the attributes of mechanical, conductive, and biological requirements are discussed (Liu et al. 2020; Saeed et al. 2023).

Chapter 6 provides conclusions and perspectives on different types of emerging nanomaterials and NPs as theranostic tools for cancer, neurodegenerative, orthopaedic, and cardiac diseases.

References

Adesoye S, Dellinger K (2022) ZnO and TiO$_2$ nanostructures for surface-enhanced Raman scattering-based biosensing: a review. Sens Bio-Sens Res 100499

Al-Jamal WT, Al-Jamal KT, Tian B, Cakebread A, Halket JM, Kostarelos K (2009) Tumor targeting of functionalized quantum dot—liposome hybrids by intravenous administration. Mol Pharm 6(2):520–530

Amirav S, Berlin S, Olszakier S, Pahari SK, Kahn I (2019) Multi-modal nano particle labeling of neurons. Front Neurosci 13:12

Ang CY, Giam L, Chan ZM, Lin AW, Gu H, Devlin E, Papaefthymiou GC, Selvan ST, Ying JY (2009) Facile synthesis of Fe$_2$O$_3$ nanocrystals without Fe (CO) 5 precursor and one-pot synthesis of highly fluorescent Fe$_2$O$_3$–CdSe nanocomposites. Adv Mater 21(8):869–873

Ang CY, Tan SY, Wang X, Zhang Q, Khan M, Bai L, Selvan ST, Ma X, Zhu L, Nguyen KT (2014) Supramolecular nanoparticle carriers self-assembled from cyclodextrin-and adamantane-functionalized polyacrylates for tumor-targeted drug delivery. J Mater Chem B 2(13):1879–1890

Ansari SR, Hempel N-J, Asad S, Svedlindh P, Bergström CA, Löbmann K, Teleki A (2022) Hyperthermia-induced in situ drug amorphization by superparamagnetic nanoparticles in oral dosage forms. ACS Appl Mater Interfaces 14(19):21978–21988

Arvanitis CD, Ferraro GB, Jain RK (2020) The blood–brain barrier and blood–tumour barrier in brain tumours and metastases. Nat Rev Cancer 20(1):26–41

Besinis A, De Peralta T, Tredwin CJ, Handy RD (2015) Review of nanomaterials in dentistry: interactions with the oral microenvironment, clinical applications, hazards, and benefits. ACS Nano 9(3):2255–2289

Bhunia SK, Saha A, Maity AR, Ray SC, Jana NR (2013) Carbon nanoparticle-based fluorescent bioimaging probes. Sci Rep 3:1473

Bui M-PN, Ahmed S, Abbas A (2015) Single-digit pathogen and attomolar detection with the naked eye using liposome-amplified plasmonic immunoassay. Nano Lett 15(9):6239–6246

Chang JC, Rosenthal SJ (2012) Visualization of lipid raft membrane compartmentalization in living RN46A neuronal cells using single quantum dot tracking. ACS Chem Neurosci 3(10):737–743

Chattrairat K, Yasui T, Suzuki S, Natsume A, Nagashima K, Iida M, Zhang M, Shimada T, Kato A, Aoki K, Ohka F, Yamazaki S, Yanagida T, Baba Y (2023) All-in-one nanowire assay system for capture and analysis of extracellular vesicles from an ex vivo brain tumor model. ACS Nano 17(3):2235–2244

Chen G, Qiu H, Prasad PN, Chen X (2014a) Upconversion nanoparticles: design, nanochemistry, and applications in theranostics. Chem Rev 114(10):5161–5214

Chen G, Tian F, Zhang Y, Zhang Y, Li C, Wang Q (2014b) Tracking of transplanted human mesenchymal stem cells in living mice using near-infrared Ag_2S quantum dots. Adv Func Mater 24(17):2481–2488

Chen H, Li B, Zhang M, Sun K, Wang Y, Peng K, Ao M, Guo Y, Gu Y (2014c) Characterization of tumor-targeting Ag_2S quantum dots for cancer imaging and therapy in vivo. Nanoscale 6(21):12580–12590

Chen O, Riedemann L, Etoc F, Herrmann H, Coppey M, Barch M, Farrar CT, Zhao J, Bruns OT, Wei H (2014d) Magneto-fluorescent core-shell supernanoparticles. Nat Commun 5:5093

Chen Y, Xu P, Shu Z, Wu M, Wang L, Zhang S, Zheng Y, Chen H, Wang J, Li Y (2014e) Multi-functional graphene oxide-based triple stimuli-responsive nanotheranostics. Adv Func Mater 24(28):4386–4396

Chen Y, Ye D, Wu M, Chen H, Zhang L, Shi J, Wang L (2014f) Break-up of two-dimensional MnO_2 nanosheets promotes ultrasensitive pH-triggered theranostics of cancer. Adv Mater 26(41):7019–7026

Chen Y, Wu Y, Sun B, Liu S, Liu H (2017) Two-dimensional nanomaterials for cancer nanotheranostics. Small 13(10):1603446

Chen J, Fan T, Xie Z, Zeng Q, Xue P, Zheng T, Chen Y, Luo X, Zhang H (2020) Advances in nanomaterials for photodynamic therapy applications: status and challenges. Biomaterials 237:119827

Cheng K, Kothapalli S-R, Liu H, Koh AL, Jokerst JV, Jiang H, Yang M, Li J, Levi J, Wu JC (2014) Construction and validation of nano gold tripods for molecular imaging of living subjects. J Am Chem Soc 136(9):3560–3571

Cheraghali S, Dini G, Caligiuri I, Back M, Rizzolio F (2023) PEG-coated MnZn ferrite nanoparticles with hierarchical structure as MRI contrast agent. Nanomaterials 13(3):452

Chieruzzi M, Pagano S, Moretti S, Pinna R, Milia E, Torre L, Eramo S (2016) Nanomaterials for tissue engineering in dentistry. Nanomaterials 6(7):134

Chithrani BD, Chan WC (2007) Elucidating the mechanism of cellular uptake and removal of protein-coated gold nanoparticles of different sizes and shapes. Nano Lett 7(6):1542–1550

Chow EK-H, Ho D (2013) Cancer nanomedicine: from drug delivery to imaging. Sci Transl Med 5(216):216rv214–216rv214

Darbandi M, Thomann R, Nann T (2005) Single quantum dots in silica spheres by microemulsion synthesis. Chem Mater 17(23):5720–5725

Das GK, Heng BC, Ng S-C, White T, Loo JSC, D'Silva L, Padmanabhan P, Bhakoo KK, Selvan ST, Tan TTY (2010) Gadolinium oxide ultranarrow nanorods as multimodal contrast agents for optical and magnetic resonance imaging. Langmuir 26(11):8959–8965

Di Mauro V, Iafisco M, Salvarani N, Vacchiano M, Carullo P, Ramírez-Rodríguez GB, Patrício T, Tampieri A, Miragoli M, Catalucci D (2016) Bioinspired negatively charged calcium phosphate nanocarriers for cardiac delivery of MicroRNAs. Nanomedicine 11(8):891–906

Diaz-Diestra D, Thapa B, Beltran-Huarac J, Weiner BR, Morell G (2017) L-cysteine capped ZnS: Mn quantum dots for room-temperature detection of dopamine with high sensitivity and selectivity. Biosens Bioelectron 87:693–700

Dinda S, Kakran M, Zeng J, Sudhaharan T, Ahmed S, Das D, Selvan ST (2016) Grafting of ZnS: Mn-doped nanocrystals and an anticancer drug onto graphene oxide for delivery and cell labeling. ChemPlusChem 81(1):100–107

Ding B, Zheng P, Ma P, Lin J (2020) Manganese oxide nanomaterials: synthesis, properties, and theranostic applications. Adv Mater 32(10):1905823

Ding Y, Ye B, Sun Z, Mao Z, Wang W (2023) Reactive oxygen species-mediated pyroptosis with the help of nanotechnology: prospects for cancer therapy. Adv NanoBiomed Res 3(1):2200077

Dou QQ, Rengaramchandran A, Selvan ST, Paulmurugan R, Zhang Y (2015) Core–shell upconversion nanoparticle–semiconductor heterostructures for photodynamic therapy. Sci Rep 5:8252

Duan S, Yang Y, Zhang C, Zhao N, Xu FJ (2017) NIR-responsive polycationic gatekeeper-cloaked hetero-nanoparticles for multimodal imaging-guided triple-combination therapy of cancer. Small 13(9):1603133

Esthar S, Rajesh J, Ayyanaar S, Kumar GGV, Thanigaivel S, Webster TJ, Rajagopal G (2023) An anti-inflammatory controlled nano drug release and pH-responsive poly lactic acid appended magnetic nanosphere for drug delivery applications. Mater Today Commun 105365

Farmanbar N, Mohseni S, Darroudi M (2022) Green synthesis of chitosan-coated magnetic nanoparticles for drug delivery of oxaliplatin and irinotecan against colorectal cancer cells. Polym Bull 79(12):10595–10613

Fawzy A, Priyadarshini B, Selvan S, Lu TB, Neo J (2017) Proanthocyanidins-loaded nanoparticles enhance dentin degradation resistance. J Dent Res 96(7):780–789

Freyer A, Sercel P, Hou Z, Savitzky BH, Kourkoutis LF, Efros AL, Krauss TD (2019) Explaining the unusual photoluminescence of semiconductor nanocrystals doped via cation exchange. Nano Lett

Gao X, Cui Y, Levenson RM, Chung LW, Nie S (2004) In vivo cancer targeting and imaging with semiconductor quantum dots. Nat Biotechnol 22(8):969

Gao X, Xu Z, Li S, Cheng L, Xu D, Li L, Chen L, Xu Y, Liu Z, Liu Y (2023) Chitosan-vancomycin hydrogel incorporated bone repair scaffold based on staggered orthogonal structure: a viable dually controlled drug delivery system. RSC Adv 13(6):3759–3765

Gerion D, Pinaud F, Williams SC, Parak WJ, Zanchet D, Weiss S, Alivisatos AP (2001) Synthesis and properties of biocompatible water-soluble silica-coated CdSe/ZnS semiconductor quantum dots. J Phys Chem B 105(37):8861–8871

Ghaffari R, Salimi-Kenari H, Fahimipour F, Rabiee SM, Adeli H, Dashtimoghadam E (2020) Fabrication and characterization of dextran/nanocrystalline β-tricalcium phosphate nanocomposite hydrogel scaffolds. Int J Biol Macromol 148:434–448

Goldman ER, Balighian ED, Mattoussi H, Kuno MK, Mauro JM, Tran PT, Anderson GP (2002) Avidin: a natural bridge for quantum dot-antibody conjugates. J Am Chem Soc 124(22):6378–6382

Goldsmith M, Abramovitz L, Peer D (2014) Precision nanomedicine in neurodegenerative diseases. ACS Nano 8(3):1958–1965

Gonçalves G, Vila M, Portolés MT, Vallet-Regi M, Gracio J, Marques PAA (2013) Nano-graphene oxide: a potential multifunctional platform for cancer therapy. Adv Healthcare Mater 2(8):1072–1090

Gong L, Yan L, Zhou R, Xie J, Wu W, Gu Z (2017) Two-dimensional transition metal dichalcogenide nanomaterials for combination cancer therapy. J Mater Chem B 5(10):1873–1895

Gu Z, Zhu S, Yan L, Zhao F, Zhao Y (2019) Graphene-based smart platforms for combined Cancer therapy. Adv Mater 31(9):1800662

Guo L, Shi Y, Liu X, Han Z, Zhao Z, Chen Y, Xie W, Li X (2018) Enhanced fluorescence detection of proteins using ZnO nanowires integrated inside microfluidic chips. Biosens Bioelectron 99:368–374

Han Q, Cai S, Yang L, Wang X, Qi C, Yang R, Wang C (2017) Molybdenum disulfide nanoparticles as multifunctional inhibitors against Alzheimer's disease. ACS Appl Mater Interfaces 9(25):21116–21123

Hanifi DA, Bronstein ND, Koscher BA, Nett Z, Swabeck JK, Takano K, Schwartzberg AM, Maserati L, Vandewal K, van de Burgt Y (2019) Redefining near-unity luminescence in quantum dots with photothermal threshold quantum yield. Science 363(6432):1199–1202

He R, Wang Y-C, Wang X, Wang Z, Liu G, Zhou W, Wen L, Li Q, Wang X, Chen X (2014) Facile synthesis of pentacle gold–copper alloy nanocrystals and their plasmonic and catalytic properties. Nat Commun 5:4327

Hong G, Robinson JT, Zhang Y, Diao S, Antaris AL, Wang Q, Dai H (2012) In vivo fluorescence imaging with Ag_2S quantum dots in the second near-infrared region. Angew Chem Int Ed 51(39):9818–9821

Hu T, Liu X, Liu S, Wang Z, Tang Z (2014) Toward understanding of transfer mechanism between electrochemiluminescent dyes and luminescent quantum dots. Anal Chem 86(8):3939–3946

Huang S, Wang L, Wang M, Zhao J, Zhang C, Ma L-Y, Jiang M, Xu L, Yu X (2023) Highly sensitive detection of extracellular vesicles on ZnO nanorods integrated microarray chips with cascade signal amplification and portable glucometer readout. Sens Actuators, B Chem 375:132878

Idris NM, Gnanasammandhan MK, Zhang J, Ho PC, Mahendran R, Zhang Y (2012) In vivo photo-dynamic therapy using upconversion nanoparticles as remote-controlled nanotransducers. Nat Med 18(10):1580

Itoo AM, Vemula SL, Gupta MT, Giram MV, Kumar SA, Ghosh B, Biswas S (2022) Multifunctional graphene oxide nanoparticles for drug delivery in cancer. J Control Release 350:26–59

Jeong H-H, Mark AG, Alarcón-Correa M, Kim I, Oswald P, Lee T-C, Fischer P (2016) Dispersion and shape engineered plasmonic nanosensors. Nat Commun 7:11331

Jiang L, Zheng R, Zeng N, Wu C, Su H (2023) In situ self-assembly of amphiphilic dextran micelles and superparamagnetic iron oxide nanoparticle-loading as magnetic resonance imaging contrast agents. Regenerative Biomater 10

Jung Y, Kim J, Kim NH, Kim HG (2023) Ag–ZnO nanocomposites as a 3D metal-enhanced fluorescence substrate for the fluorescence detection of DNA. ACS Appl Nano Mater

Kamaly N, He JC, Ausiello DA, Farokhzad OC (2016) Nanomedicines for renal disease: current status and future applications. Nat Rev Nephrol 12(12):738

Kang T, Li F, Baik S, Shao W, Ling D, Hyeon T (2017) Surface design of magnetic nanoparticles for stimuli-responsive cancer imaging and therapy. Biomaterials 136:98–114

Karageorgou M-A, Bouziotis P, Stiliaris E, Stamopoulos D (2023) Radiolabeled iron oxide nanoparticles as dual modality contrast agents in SPECT/MRI and PET/MRI. Nanomaterials 13(3):503

Kleinstreuer C, Chari SV, Vachhani S (2018) Potential use of multifunctional nanoparticles for the treatment of cardiovascular diseases. J Cardiol Cardiovasc Sci 2(3)

Kong Y, Chen J, Fang H, Heath G, Wo Y, Wang W, Li Y, Guo Y, Evans SD, Chen S (2016) Highly fluorescent ribonuclease-A-encapsulated lead sulfide quantum dots for ultrasensitive fluorescence in vivo imaging in the second near-infrared window. Chem Mater 28(9):3041–3050

Kosaka PM, Pini V, Ruz JJ, Da Silva R, González M, Ramos D, Calleja M, Tamayo J (2014) Detection of cancer biomarkers in serum using a hybrid mechanical and optoplasmonic nanosensor. Nat Nanotechnol 9(12):1047

Lane LA, Qian X, Nie S (2015) SERS nanoparticles in medicine: from label-free detection to spectroscopic tagging. Chem Rev 115(19):10489–10529

Langer J, Novikov SM, Liz-Marzán LM (2015) Sensing using plasmonic nanostructures and nanoparticles. Nanotechnology 26(32):322001

Le Floc'h J, Lu HD, Lim TL, Démoré C, Prud'homme RK, Hynynen K, Foster FS (2019) Transcranial photoacoustic detection of blood-brain barrier disruption following focused ultrasound-mediated nanoparticle delivery. Mol Imaging Biol 1–11

Lee D-E, Koo H, Sun I-C, Ryu JH, Kim K, Kwon IC (2012) Multifunctional nanoparticles for multimodal imaging and theragnosis. Chem Soc Rev 41(7):2656–2672

Li X, Chen L (2016) Fluorescence probe based on an amino-functionalized fluorescent magnetic nanocomposite for detection of folic acid in serum. ACS Appl Mater Interfaces 8(46):31832–31840

Liu T, Wang C, Gu X, Gong H, Cheng L, Shi X, Feng L, Sun B, Liu Z (2014) Drug delivery with PEGylated MoS_2 nano-sheets for combined photothermal and chemotherapy of cancer. Adv Mater 26(21):3433–3440

Liu B, Li C, Chen Y, Zhang Y, Hou Z, Huang S, Lin J (2015) Multifunctional $NaYF_4$: Yb, Er@ $mSiO_2$ @ Fe_3O_4-PEG nanoparticles for UCL/MR bioimaging and magnetically targeted drug delivery. Nanoscale 7(5):1839–1848

Liu T-M, Conde J, Lipiński T, Bednarkiewicz A, Huang C-C (2017) Smart NIR linear and nonlinear optical nanomaterials for cancer theranostics: prospects in photomedicine. Prog Mater Sci 88:89–135

Liu X, Zhang W, Huang L, Hu N, Liu W, Liu Y, Li S, Yang C, Suo Y, Wang J (2018) Fluorometric determination of dopamine by using molybdenum disulfide quantum dots. Microchim Acta 185(4):234

Liu X-G, Sun Y-Q, Bian J, Han T, Yue D-D, Li D-Q, Gao P-Y (2019a) Neuroprotective effects of triterpenoid saponins from Medicago sativa L. against H2O2-induced oxidative stress in SH-SY5Y cells. Bioorg Chem 83:468–476

Liu Y, Meng X, Bu W (2019b) Upconversion-based photodynamic cancer therapy. Coord Chem Rev 379:82–98

Liu S, Chen X, Bao L, Liu T, Yuan P, Yang X, Qiu X, Gooding JJ, Bai Y, Xiao J (2020) Treatment of infarcted heart tissue via the capture and local delivery of circulating exosomes through antibody-conjugated magnetic nanoparticles. Nat Biomed Eng 4(11):1063–1075

Liu X, Zhang H, Zhang T, Wang Y, Jiao W, Lu X, Gao X, Xie M, Shan Q, Wen N (2021) Magnetic nanomaterials-mediated cancer diagnosis and therapy. Prog Biomed Eng 4(1):012005

Liu Z, Wang X, Chen X, Cui L, Li Z, Bai Z, Lin K, Yang J, Tian F (2023) Construction of pH-responsive polydopamine coated magnetic layered hydroxide nanostructure for intracellular drug delivery. Eur J Pharm Biopharm 182:12–20

Loch RA, Wang H, Marín AP, Berger P, Nielsen H, Chroni A, Luo J (2023) Cross interactions between apolipoprotein E and amyloid proteins in neurodegenerative diseases. Comput Struct Biotechnol J

Lu H, Li W, Qiu P, Zhang X, Qin J, Cai Y, Lu X (2022) MnO$_2$ doped graphene nanosheets for carotid body tumor combination therapy. Nanoscale Adv 4(20):4304–4313

Marom O, Nakhoul F, Tisch U, Shiban A, Abassi Z, Haick H (2012) Gold nanoparticle sensors for detecting chronic kidney disease and disease progression. Nanomedicine 7(5):639–650

Mattoussi H, Mauro JM, Goldman ER, Anderson GP, Sundar VC, Mikulec FV, Bawendi MG (2000) Self-assembly of CdSe−ZnS quantum dot bioconjugates using an engineered recombinant protein. J Am Chem Soc 122(49):12142–12150

Mazaheri M, Eslahi N, Ordikhani F, Tamjid E, Simchi A (2015) Nanomedicine applications in orthopedic medicine: state of the art. Int J Nanomed 10:6039

Medintz IL, Uyeda HT, Goldman ER, Mattoussi H (2005) Quantum dot bioconjugates for imaging, labelling and sensing. Nat Mater 4(6):435

Min Y, Caster JM, Eblan MJ, Wang AZ (2015) Clinical translation of nanomedicine. Chem Rev 115(19):11147–11190

Mohapatra J, Nigam S, George J, Arellano AC, Wang P, Liu JP (2023) Principles and applications of magnetic nanomaterials in magnetically guided bioimaging. Mater Today Phys 101003

Mukherjee S, Liang L, Veiseh O (2020) Recent advancements of magnetic nanomaterials in cancer therapy. Pharmaceutics 12(2):147

Mulvaney P, Liz-Marzan L, Giersig M, Ung T (2000) Silica encapsulation of quantum dots and metal clusters. J Mater Chem 10(6):1259–1270

Murcia MJ, Minner DE, Mustata G-M, Ritchie K, Naumann CA (2008) Design of quantum dot-conjugated lipids for long-term, high-speed tracking experiments on cell surfaces. J Am Chem Soc 130(45):15054–15062

Murugan C, Park S (2023) Cerium ferrite@ molybdenum disulfide nanozyme for intracellular ROS generation and photothermal-based cancer therapy. J Photochem Photobiol, A 437:114466

Naik J, David M (2023) ROS mediated apoptosis and cell cycle arrest in human lung adenocarcinoma cell line by silver nanoparticles synthesized using Swietenia macrophylla seed extract. J Drug Deliv Sci Technol 80:104084

Nam J, Son S, Park KS, Zou W, Shea LD, Moon JJ (2019) Cancer nanomedicine for combination cancer immunotherapy. Nat Rev Mater 1

Nanda SS, Kim MJ, Kim K, Papaefthymiou GC, Selvan ST, Yi DK (2017) Recent advances in biocompatible semiconductor nanocrystals for immunobiological applications. Colloids Surf, B 159:644–654

Narayanan K, Yen SK, Dou Q, Padmanabhan P, Sudhaharan T, Ahmed S, Ying JY, Selvan ST (2013) Mimicking cellular transport mechanism in stem cells through endosomal escape of new peptide-coated quantum dots. Sci Rep 3:2184

Nguyen KT, Zhao Y (2015) Engineered hybrid nanoparticles for on-demand diagnostics and therapeutics. Acc Chem Res 48(12):3016–3025

Ondry JC, Philbin JP, Lostica M, Rabani E, Alivisatos AP (2019) Resilient pathways to atomic attachment of quantum dot dimers and artificial solids from faceted CdSe quantum dot building blocks. ACS Nano

Ouyang R, Zhang Q, Cao P, Yang Y, Zhao Y, Liu B, Miao Y, Zhou S (2023) Efficient improvement in chemo/photothermal synergistic therapy against lung cancer using Bi@ Au nano-acanthospheres. Colloids Surf, B 222:113116

PA Ferreira M, Balasubramanian V, Hirvonen J, Ruskoaho H, Santos HA (2015) Advanced nanomedicines for the treatment and diagnosis of myocardial infarction and heart failure. Curr Drug Targets 16(14):1682–1697

Padovani GC, Feitosa VP, Sauro S, Tay FR, Durán G, Paula AJ, Durán N (2015) Advances in dental materials through nanotechnology: facts, perspectives and toxicological aspects. Trends Biotechnol 33(11):621–636

Paisrisarn P, Yasui T, Zhu Z, Klamchuen A, Kasamechonchung P, Wutikhun T, Yordsri V, Baba Y (2022) Tailoring ZnO nanowire crystallinity and morphology for label-free capturing of extracellular vesicles. Nanoscale 14(12):4484–4494

Palui G, Aldeek F, Wang W, Mattoussi H (2015) Strategies for interfacing inorganic nanocrystals with biological systems based on polymer-coating. Chem Soc Rev 44(1):193–227

Peng Z, Han X, Li S, Al-Youbi AO, Bashammakh AS, El-Shahawi MS, Leblanc RM (2017) Carbon dots: biomacromolecule interaction, bioimaging and nanomedicine. Coord Chem Rev 343:256–277

Perli M, Karagkiozaki V, Pappa F, Moutsios I, Tzounis L, Zachariadis A, Gravalidis C, Laskarakis A, Logothetidis S (2017) Synthesis and characterization of Ag nanoparticles for orthopaedic applications. Mater Today: Proc 4(7):6889–6900

Priyadarshini B, Selvan S, Lu T, Xie H, Neo J, Fawzy A (2016) Chlorhexidine nanocapsule drug delivery approach to the resin-dentin interface. J Dent Res 95(9):1065–1072

Priyadarshini B, Selvan S, Narayanan K, Fawzy A (2017) Characterization of chlorhexidine-loaded calcium-hydroxide microparticles as a potential dental pulp-capping material. Bioengineering 4(3):59

Ramanathan S, Archunan G, Sivakumar M, Selvan ST, Fred AL, Kumar S, Gulyás B, Padmanabhan P (2018) Theranostic applications of nanoparticles in neurodegenerative disorders. Int J Nanomed 13:5561

Ray S, Cheng C-A, Chen W, Li Z, Zink JI, Lin Y-Y (2019) Magnetic heating stimulated cargo release with dose control using multifunctional MR and thermosensitive liposome. Nanotheranostics 3(2):166

Ren H-B, Wu B-Y, Chen J-T, Yan X-P (2011) Silica-coated S_2—enriched manganese-doped ZnS quantum dots as a photoluminescence probe for imaging intracellular Zn^{2+} ions. Anal Chem 83(21):8239–8244

Saeed S, Khan SU, Gul R (2023) Nanoparticle: a promising player in nanomedicine and its theranostic applications for the treatment of cardiovascular diseases. Curr Probl Cardiol 101599

Saraiva C, Praça C, Ferreira R, Santos T, Ferreira L, Bernardino L (2016) Nanoparticle-mediated brain drug delivery: overcoming blood–brain barrier to treat neurodegenerative diseases. J Control Release 235:34–47

Sasaki A, Tsukasaki Y, Komatsuzaki A, Sakata T, Yasuda H, Jin T (2015) Recombinant protein (EGFP-Protein G)-coated PbS quantum dots for in vitro and in vivo dual fluorescence (visible and second-NIR) imaging of breast tumors. Nanoscale 7(12):5115–5119

Satpathy M, Wang L, Zielinski RJ, Qian W, Wang YA, Mohs AM, Kairdolf BA, Ji X, Capala J, Lipowska M (2019) Targeted drug delivery and image-guided therapy of heterogeneous ovarian cancer using HER2-targeted theranostic nanoparticles. Theranostics 9(3):778

Scaletti F, Hardie J, Lee Y-W, Luther DC, Ray M, Rotello VM (2018) Protein delivery into cells using inorganic nanoparticle–protein supramolecular assemblies. Chem Soc Rev 47(10):3421–3432

Selvan ST (2010) Silica-coated quantum dots and magnetic nanoparticles for bioimaging applications (mini-review). Biointerphases 5(3) FA110–FA115

Selvan ST, Li C, Ando M, Murase N (2004) Formation of luminescent CdTe–silica nanoparticles through an inverse microemulsion technique. Chem Lett 33(4):434–435

Selvan ST, Tan TT, Ying JY (2005) Robust, non-cytotoxic, silica-coated CdSe quantum dots with efficient photoluminescence. Adv Mater 17(13):1620–1625

Selvan ST, Patra PK, Ang CY, Ying JY (2007) Synthesis of silica-coated semiconductor and magnetic quantum dots and their use in the imaging of live cells. Angew Chem Int Ed 46(14):2448–2452

Selvan ST, Tan TTY, Yi DK, Jana NR (2009) Functional and multifunctional nanoparticles for bioimaging and biosensing. Langmuir 26(14):11631–11641

Seyedsalehi A, Daneshmandi L, Barajaa M, Riordan J, Laurencin CT (2020) Fabrication and characterization of mechanically competent 3D printed polycaprolactone-reduced graphene oxide scaffolds. Sci Rep 10(1):22210

Shabalkin ID, Komlev AS, Tsymbal SA, Burmistrov OI, Zverev VI, Krivoshapkin PV (2023) Multifunctional tunable $ZnFe_2O_4$ @ $MnFe_2O_4$ nanoparticles for dual-mode MRI and combined magnetic hyperthermia with radiotherapy treatment. J Mater Chem B

Shi Y, Pan Y, Zhong J, Yang J, Zheng J, Cheng J, Song R, Yi C (2015) Facile synthesis of gadolinium (III) chelates functionalized carbon quantum dots for fluorescence and magnetic resonance dual-modal bioimaging. Carbon 93:742–750

Shukla R, Singh A, Handa M, Flora S, Kesharwani P (2021) Nanotechnological approaches for targeting amyloid-β aggregation with potential for neurodegenerative disease therapy and diagnosis. Drug Discovery Today 26(8):1972–1979

Snyder EL, Bailey D, Shipitsin M, Polyak K, Loda M (2009) Identification of CD44v6+/CD24— breast carcinoma cells in primary human tumors by quantum dot-conjugated antibodies. Lab Invest 89(8):857

Soh M, Kang DW, Jeong HG, Kim D, Kim DY, Yang W, Song C, Baik S, Choi IY, Ki SK (2017) Ceria–zirconia nanoparticles as an enhanced multi-antioxidant for sepsis treatment. Angew Chem Int Ed 56(38):11399–11403

Song G, Hao J, Liang C, Liu T, Gao M, Cheng L, Hu J, Liu Z (2016) Degradable molybdenum oxide nanosheets with rapid clearance and efficient tumor homing capabilities as a therapeutic nanoplatform. Angew Chem Int Ed 55(6):2122–2126

Su Y, Yang F, Wang M, Cheung PC (2023) Cancer immunotherapeutic effect of carboxymethylated β-d-glucan coupled with iron oxide nanoparticles via reprogramming tumor-associated macrophages. Int J Biol Macromol 228:692–705

Tamil Selvan S, Ravichandar R, Kanta Ghosh K, Mohan A, Mahalakshmi P, Gulyás B, Padmanabhan P (2021) Coordination chemistry of ligands: insights into the design of amyloid beta/tau-PET imaging probes and nanoparticles-based therapies for Alzheimer's disease. Coord Chem Rev 430:213659

Tan TT, Selvan ST, Zhao L, Gao S, Ying JY (2007) Size control, shape evolution, and silica coating of near-infrared-emitting PbSe quantum dots. Chem Mater 19(13):3112–3117

Tan C, Cao X, Wu X-J, He Q, Yang J, Zhang X, Chen J, Zhao W, Han S, Nam G-H (2017) Recent advances in ultrathin two-dimensional nanomaterials. Chem Rev 117(9):6225–6331

Tapeinos C, Battaglini M, Ciofani G (2017) Advances in the design of solid lipid nanoparticles and nanostructured lipid carriers for targeting brain diseases. J Control Release 264:306–332

Tay CY, Setyawati MI, Xie J, Parak WJ, Leong DT (2014) Back to basics: exploiting the innate physico-chemical characteristics of nanomaterials for biomedical applications. Adv Func Mater 24(38):5936–5955

Teleanu DM, Chircov C, Grumezescu AM, Volceanov A, Teleanu RI (2019a) Contrast agents delivery: an up-to-date review of nanodiagnostics in neuroimaging. Nanomaterials 9(4):542

Teleanu DM, Negut I, Grumezescu V, Grumezescu AM, Teleanu RI (2019b) Nanomaterials for drug delivery to the central nervous system. Nanomaterials 9(3):371

Tian B, Al-Jamal KT, Kostarelos K (2011) Doxorubicin-loaded lipid-quantum dot hybrids: surface topography and release properties. Int J Pharm 416(2):443–447

Topete A, Alatorre-Meda M, Villar-Alvarez EM, Carregal-Romero S, Barbosa S, Parak WJ, Taboada P, Mosquera V (2014) Polymeric-gold nanohybrids for combined imaging and cancer therapy. Adv Healthcare Mater 3(8):1309–1325

Turrina C, Milani D, Klassen A, Rojas-González DM, Cookman J, Opel M, Sartori B, Mela P, Berensmeier S, Schwaminger SP (2022) Carboxymethyl-dextran-coated superparamagnetic iron oxide nanoparticles for drug delivery: influence of the coating thickness on the particle properties. Int J Mol Sci 23(23):14743

Vangijzegem T, Lecomte V, Ternad I, Van Leuven L, Muller RN, Stanicki D, Laurent S (2023) Superparamagnetic iron oxide nanoparticles (SPION): from fundamentals to state-of-the-art innovative applications for cancer therapy. Pharmaceutics 15(1):236

Wang Q, Bao Y, Ahire J, Chao Y (2013) Co-encapsulation of biodegradable nanoparticles with silicon quantum dots and quercetin for monitored delivery. Adv Healthcare Mater 2(3):459–466

Wang S, Li X, Chen Y, Cai X, Yao H, Gao W, Zheng Y, An X, Shi J, Chen H (2015a) A facile one-pot synthesis of a two-dimensional MoS_2/Bi_2S_3 composite theranostic nanosystem for multi-modality tumor imaging and therapy. Adv Mater 27(17):2775–2782

Wang Y, Song S, Liu J, Liu D, Zhang H (2015b) ZnO-functionalized upconverting nanotheranostic agent: multi-modality imaging-guided chemotherapy with on-demand drug release triggered by pH. Angew Chem Int Ed 54(2):536–540

Wang Z, Chang Z, Lu M, Shao D, Yue J, Yang D, Zheng X, Li M, He K, Zhang M (2018) Shape-controlled magnetic mesoporous silica nanoparticles for magnetically-mediated suicide gene therapy of hepatocellular carcinoma. Biomaterials 154:147–157

Wang J, Xu M, Wang K, Chen Z (2019) Stable mesoporous silica nanoparticles incorporated with MoS_2 and AIE for targeted fluorescence imaging and photothermal therapy of cancer cells. Colloids Surf, B 174:324–332

Wang Q, Li M, Cui T, Wu R, Guo F, Fu M, Zhu Y, Yang C, Chen B, Sun G (2023a) A novel Zwitterionic hydrogel incorporated with graphene oxide for bone tissue engineering: synthesis, characterization, and promotion of osteogenic differentiation of bone mesenchymal stem cells. Int J Mol Sci 24(3):2691

Wang Y, Gong F, Han Z, Lei H, Zhou Y, Cheng S, Yang X, Wang T, Wang L, Yang N (2023b) Oxygen-deficient molybdenum oxide nanosensitizers for ultrasound-enhanced cancer metalloimmunotherapy. Angew Chem

Wang Z, Ren X, Wang D, Guan L, Li X, Zhao Y, Liu A, He L, Wang T, Zvyagin AV (2023c) Novel strategies for tumor radiosensitization mediated by multifunctional gold-based nanomaterials. Biomater Sci

Weng KC, Noble CO, Papahadjopoulos-Sternberg B, Chen FF, Drummond DC, Kirpotin DB, Wang D, Hom YK, Hann B, Park JW (2008) Targeted tumor cell internalization and imaging of multifunctional quantum dot-conjugated immunoliposomes in vitro and in vivo. Nano Lett 8(9):2851–2857

Williams RM, Jaimes EA, Heller DA (2016) Nanomedicines for kidney diseases. Kidney Int 90(4):740–745

Wu P, Zhao T, Tian Y, Wu L, Hou X (2013) Protein-directed synthesis of Mn-doped ZnS quantum dots: a dual-channel biosensor for two proteins. Chem—Eur J 19(23):7473–7479

Wu P, Zhang J, Wang S, Zhu A, Hou X (2014) Sensing during in situ growth of Mn-doped ZnS QDs: a phosphorescent sensor for detection of H_2S in biological samples. Chem—Eur J 20(4):952–956

Wu Q, Chen L, Huang L, Wang J, Liu J, Hu C, Han H (2015) Quantum dots decorated gold nanorod as fluorescent-plasmonic dual-modal contrasts agent for cancer imaging. Biosens Bioelectron 74:16–23

Wu M, Xue Y, Li N, Zhao H, Lei B, Wang M, Wang J, Luo M, Zhang C, Du Y (2019) Tumor-microenvironment-induced degradation of ultrathin gadolinium oxide nanoscrolls for magnetic-resonance-imaging-monitored, activatable cancer chemotherapy. Angew Chem 131(21):6954–6959

Xie T, Jing C, Long Y-T (2017) Single plasmonic nanoparticles as ultrasensitive sensors. Analyst 142(3):409–420

Xing P, Zhao Y (2016) Multifunctional nanoparticles self-assembled from small organic building blocks for biomedicine. Adv Mater 28(34):7304–7339

Xu Z-Q, Lan J-Y, Jin J-C, Dong P, Jiang F-L, Liu Y (2015) Highly photoluminescent nitrogen-doped carbon nanodots and their protective effects against oxidative stress on cells. ACS Appl Mater Interfaces 7(51):28346–28352

Xu G, Zeng S, Zhang B, Swihart MT, Yong K-T, Prasad PN (2016) New generation cadmium-free quantum dots for biophotonics and nanomedicine. Chem Rev 116(19):12234–12327

Xu J, Xu L, Wang C, Yang R, Zhuang Q, Han X, Dong Z, Zhu W, Peng R, Liu Z (2017) Near-infrared-triggered photodynamic therapy with multitasking upconversion nanoparticles in combination with checkpoint blockade for immunotherapy of colorectal cancer. ACS Nano 11(5):4463–4474

Yan Y, Gong J, Chen J, Zeng Z, Huang W, Pu K, Liu J, Chen P (2019) Recent advances on graphene quantum dots: from chemistry and physics to applications. Adv Mater 31(21):1808283

Yang H-Y, Zhao Y-W, Zhang Z-Y, Xiong H-M, Yu S-N (2013) One-pot synthesis of water-dispersible Ag_2S quantum dots with bright fluorescent emission in the second near-infrared window. Nanotechnology 24(5):055706

Yang K, Feng L, Liu Z (2016a) Stimuli responsive drug delivery systems based on nano-graphene for cancer therapy. Adv Drug Deliv Rev 105:228–241

Yang S, Dai X, Stogin BB, Wong T-S (2016b) Ultrasensitive surface-enhanced Raman scattering detection in common fluids. Proc Natl Acad Sci 113(2):268–273

Yang W, Guo W, Chang J, Zhang B (2017) Protein/peptide-templated biomimetic synthesis of inorganic nanoparticles for biomedical applications. J Mater Chem B 5(3):401–417

Yang HY, Li Y, Lee DS (2018) Multifunctional and stimuli-responsive magnetic nanoparticle-based delivery systems for biomedical applications. Adv Therap 1(2)

Yang B, Chen Y, Shi J (2019) Reactive oxygen species (ROS)-based nanomedicine. Chem Rev 119(8):4881–4985

Yao X, Tian Z, Liu J, Zhu Y, Hanagata N (2016) Mesoporous silica nanoparticles capped with Graphene quantum dots for potential chemo—photothermal synergistic Cancer therapy. Langmuir 33(2):591–599

Yen SK, Selvan ST (2015) Fluorescence retrieval of CdSe quantum dots by self-assembly of supramolecular aggregates of reverse micelles. Small 11(22):2619

Yen SK, Janczewski D, Lakshmi JL, Dolmanan SB, Tripathy S, Ho VH, Vijayaragavan V, Hariharan A, Padmanabhan P, Bhakoo KK (2013a) Design and synthesis of polymer-functionalized NIR fluorescent dyes–magnetic nanoparticles for bioimaging. ACS Nano 7(8):6796–6805

Yen SK, Padmanabhan P, Selvan ST (2013b) Multifunctional iron oxide nanoparticles for diagnostics, therapy and macromolecule delivery. Theranostics 3(12):986

Yen SK, Varma DP, Guo WM, Ho VH, Vijayaragavan V, Padmanabhan P, Bhakoo K, Selvan ST (2015) Synthesis of small-sized, porous, and low-toxic magnetite nanoparticles by thin POSS silica coating. Chem—Eur J 21(10):3914–3918

Yi DK, Selvan ST, Lee SS, Papaefthymiou GC, Kundaliya D, Ying JY (2005) Silica-coated nanocomposites of magnetic nanoparticles and quantum dots. J Am Chem Soc 127(14):4990–4991

Yi DK, Nanda SS, Kim K, Selvan ST (2017) Recent progress in nanotechnology for stem cell differentiation, labeling, tracking and therapy. J Mater Chem B 5(48):9429–9451

Zhang M, Bai L, Shang W, Xie W, Ma H, Fu Y, Fang D, Sun H, Fan L, Han M (2012) Facile synthesis of water-soluble, highly fluorescent graphene quantum dots as a robust biological label for stem cells. J Mater Chem 22(15):7461–7467

Zhang Y, Das GK, Vijayaragavan V, Xu QC, Padmanabhan P, Bhakoo KK, Selvan ST, Tan TTY (2014) "Smart" theranostic lanthanide nanoprobes with simultaneous up-conversion fluorescence and tunable T1–T2 magnetic resonance imaging contrast and near-infrared activated photodynamic therapy. Nanoscale 6(21):12609–12617

Zhang S, Geryak R, Geldmeier J, Kim S, Tsukruk VV (2017) Synthesis, assembly, and applications of hybrid nanostructures for biosensing. Chem Rev 117(20):12942–13038

Zhang P, Cui Y, Anderson CF, Zhang C, Li Y, Wang R, Cui H (2018) Peptide-based nanoprobes for molecular imaging and disease diagnostics. Chem Soc Rev 47(10):3490–3529

Zhang X, Guo Z, Zhang X, Gong L, Dong X, Fu Y, Wang Q, Gu Z (2019) Mass production of poly (ethylene glycol) monooleate-modified core-shell structured upconversion nanoparticles for bio-imaging and photodynamic therapy. Sci Rep 9(1):5212

Zhelev Z, Ohba H, Bakalova R (2006) Single quantum dot-micelles coated with silica shell as potentially non-cytotoxic fluorescent cell tracers. J Am Chem Soc 128(19):6324–6325

Zheng F-F, Zhang P-H, Xi Y, Chen J-J, Li L-L, Zhu J-J (2015a) Aptamer/graphene quantum dots nanocomposite capped fluorescent mesoporous silica nanoparticles for intracellular drug delivery and real-time monitoring of drug release. Anal Chem 87(23):11739–11745

Zheng XT, Ananthanarayanan A, Luo KQ, Chen P (2015b) Glowing graphene quantum dots and carbon dots: properties, syntheses, and biological applications. Small 11(14):1620–1636

Zhou B, Shi B, Jin D, Liu X (2015) Controlling upconversion nanocrystals for emerging applications. Nat Nanotechnol 10(11):924

Zhou Z, Song J, Nie L, Chen X (2016) Reactive oxygen species generating systems meeting challenges of photodynamic cancer therapy. Chem Soc Rev 45(23):6597–6626

Chapter 2
Cancer Nanomedicine

Abstract This chapter deals with emerging QDs and NPs that can be utilized as bioimaging probes for cancer nanomedicine. It covers topics on deep-tissue bimodal imaging utilizing NIR emitting QDs and other interesting Au-NPs and tripods for bioimaging and therapy. Drug carriers such as iron oxide NPs, silica-, graphene-, graphene oxide- based nanocarriers, peptides or polymer mediated delivery vehicles, and liposomes for targeted delivery are discussed. Metallic nanostructures and plasmonic bimetallic nanocrystals and core–shell nanocomposites for combined photothermal-/chemo- or radio-/chemo-therapy are delineated.

Keywords Cancer · Bioimaging · Therapy · Nanoparticles · Nanomedicine · Targeted delivery · Diagnostics · Photothermal therapy

Cancer is one of the leading causes of human death throughout the globe. Toward combating cancer, tremendous efforts put forward in recent years in drug discovery and pharmacological research. However, a complete cure is beyond the reality. As a thumb rule, prevention is better than cure. Despite the fact that the Food and Drug Administration (FDA) have approved several drugs, cisplatin and derivatives are still considered as one of the promising therapeutic strategies for cancer (Ruiz-Ceja and Chirino 2017). Although the drugs are capable of targeting the epidermal growth factor receptor (EGFR), none of the pharmacological treatments succeeded in the complete cure of cancer even in combination with radiation and surgery. Therefore, the early detection of cancer biomarkers is mandatory, for which nanotechnology is an interesting approach. Here, a plethora of questions arises: How do we achieve early diagnosis? How do we deliver approved drugs more specifically to cancer cells? What are their side effects? Would they affect normal, non-cancerous cells?

In drug discovery and pharmacological research, thiazolo[4,5-d]pyrimidines emerge as immune-modulators, anticancer, anti-Parkinson's, antibacterial, antiviral, and antifungal agents (Kuppast and Fahmy 2016). Furthermore, several drug candidates such as pyrazole based compounds with lesser side effects and improved efficacy are being explored (Ganguly and Jacob 2017). Microtubule-stabilizing agents/drugs have attracted interest in clinical drug discovery, cancer therapy studies (Zhao et al. 2016), and neurodegenerative diseases (Brunden et al. 2017).

© The Author(s), under exclusive license to Springer Nature Singapore Pte Ltd. 2023   17
S. Tamil Selvan, *Nanomedicine*, Nanotheranostics,
https://doi.org/10.1007/978-981-99-2139-3_2

The interplay between nanotechnology and medicinal chemistry has created new avenues in the nanocarrier formulations (Sunoqrot et al. 2017). Various drug delivery vehicles based on nanocarrier formulations such as polymeric micelles (Nasongkla et al. 2006; Gong et al. 2012; Kataoka et al. 2012), liposomes (Hwang et al. 2016; Malinge et al. 2017; Man et al. 2019), and polymer/inorganic NPs for anticancer drug/siRNA delivery (Zhang et al. 2011; Meng et al. 2013; Qu et al. 2019) have been explored to increase the drug solubility, and bioavailability.

Cancer nanomedicine has advanced significantly in recent years (Ali et al. 2017; Komatsu 2023). Molecular imaging has advanced cancer nanomedicine by integrating different multimodality techniques such as MRI, near-infrared optical imaging, positron emission topography (PET), computed tomography (CT), etc. (Pratt et al. 2016; Dearling and Packard 2017; Liu et al. 2017a; Miller et al. 2017).

2.1 Emerging QDs and NPs as Bioimaging Probes

Both CdSe/ZnS—based QDs and iron oxide (Fe_3O_4 or Fe_2O_3) NPs dominated the bioimaging arena especially in optical and MR imaging, respectively. Although CdSe/ZnS QDs have been used extensively as cell labelling probes, the presence of heavy metals poses issues related to toxicity.

This section deals with the emergence of silver based QDs (Ag_2S, Ag_2Se), up-conversion NPs (UCNPs), Au-NPs, and their fabrication as either bimodal or trimodal agents for biosensing and imaging.

2.1.1 NIR Emitting QDs for Deep-Tissue Bimodal Imaging

Bioimaging probes in the near-infrared window II (NIR-II, 1000–1700 nm) are highly promising for in vivo imaging with enhanced resolution and deeper tissue penetration. Earlier, Ag_2Se QDs synthesized in the size regime of 2–3 nm exhibited the emission wavelengths in the range of 700–820 nm, depending on the size (Gu et al. 2011). This work also elucidated the promise of these less toxic NIR emitting Ag_2Se QDs for deep-tissue in vivo imaging of a nude mouse after abdominal injection and detection on its backside.

In another interesting work, similar Ag_2Se QDs have demonstrated their potentials in trimodality imaging, encompassing fluorescence, MRI, and PET (Tian et al. 2019). This work further suggested that a high tumor-to-muscle ratio of nine in PET imaging achieved after conjugation of the particles with a targeting peptide. Finally, the QDs excreted within 12 h from the body by the kidneys. Alternatively, Ag_2S QDs can be used as sensors for fluoride ions detection in living cells (Hong et al. 2012; Hong et al. 2017). These NIR emitting Ag_2S QDs (emission at 795 nm) showed good sensitivity for F^- detection with a detection limit of 1.5 μM (Ding et al. 2017).

Very recently multifunctional NPs combining photoluminescent PbS/CdS QDs (emitting in the second biological window of 1000–1350 nm; 14 mm deep-tissue imaging) and superparamagnetic Fe_3O_4 NPs (higher T2 relaxivity, 282 mM^{-1} s^{-1}) have been developed for in vivo bimodal imaging (MR and optical), and bimodal therapy (based on magneto-, and photo-thermal heating) (Yang et al. 2019).

2.1.2 Au-NPs and Tripods for Bioimaging and Therapy

Au-NPs can act as QDs when they exist in the form of clusters (Yahia-Ammar et al. 2016; Khandelwal and Poddar 2017). Such Au_{22} clusters were synthesized, thanks to the facile thiol conjugation chemistry (Yu et al. 2014; Pyo et al. 2015). Ultra-bright Au_{22} clusters functionalized with thiolate ligands such as glutathione yielded high photoluminescence (PL) quantum yield (QY) (> 60%) in toluene after rigidifying the Au-shell with tetraoctylammonium (TOA) cations. However, transferring the clusters back to water decreased the PL-QY to < 10. This method warranted a facile phase transfer approach to retain the high PL-QY. Interestingly, the same group reported folate-functionalized Au_{22} clusters (Au_{22}-FA) with a PL-QY of 42%, enabling them for imaging of HeLa cancer cells (Pyo et al. 2017). Essentially, gold nanoclusters (Au NCs) provide several key features such as non-toxicity, high renal clearance (Chen et al. 2016), passive tumor targeting (Liu et al. 2013), light induced cell death (Zhang et al. 2015), drug delivery carriers (Yahia-Ammar et al. 2016), and optical sensing and biodetection (Shang et al. 2013), making them more useful for cancer nanomedicine.

Earlier, Xing et al. developed an intriguing trimodal nanoprobe, consisting of up-conversion NPs, and Au-NPs encapsulated within silica ($NaY(Gd)F_4:Yb^{3+}/Er^{3+}/Tm^{3+}@SiO_2$-Au) for multimodality imaging (fluorescence, MR and CT) of tumor bearing mice in vivo (Xing et al. 2012). This approach enabled for ameliorating contrast effects necessary for in vivo optical, MR and CT imaging, owing to the presence of up-conversion NPs for optical, Gd ions for T1 MR contrast, and Au NPs for CT.

Nano Au tripods have been studied for molecular imaging of living subjects, utilizing positron emission tomography (PET) and photoacoustic (PA) imaging modalities (Cheng et al. 2014). We have recently developed Cu–Au tripods for photothermal anticancer therapy (Nanda et al. 2019) Some of the typical examples of NPs serving as bioimaging probes are shown in Fig. 2.1. In our group, bifunctional NPs (MNP@Dye-Pol) consisting of magnetic iron oxide NPs (MNPs) and IR-820 dye functionalized with an amphiphilic polymer, poly-(isobutylene-alt-maleic anhydride) for NIR cell imaging and MR animal imaging have been developed (Fig. 2.1a) (Yen et al. 2013). Conversely, tripod Au NPs have been used for PET and PA animal imaging (Fig. 2.1b) (Cheng et al. 2014). Therapeutic applications employing NPs of different shapes (e.g. tripods, (Nanda et al. 2019) and nanoprisms (Pérez-Hernández

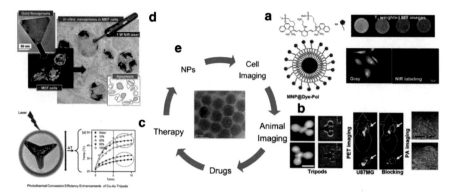

Fig. 2.1 Schematic representation of nanoparticles (NPs) used in cell labelling, animal imaging, and therapeutic application. **a** Bimodality imaging of NIR-dye@polymer—conjugated magnetic NPs for optical and magnetic resonance imaging. Reproduced with permission (Yen et al. 2013). Copyright 2013, ACS. **b** Au-tripod for PET and PA imaging. Reproduced with permission (Cheng et al. 2014). Copyright 2014, ACS. **c** Photothermal conversion efficiency endowed by Cu–Au tripod nanocrystals. Reproduced with permission (Nanda et al. 2019). Copyright 2019, ACS. **d** Au-nanoprisms for NIR-assisted cancer therapy. Reproduced with permission (Pérez-Hernández et al. 2015). Copyright 2015, ACS. **e** A cartoon depicting the interplay of NPs from cell/animal imaging to therapy. Inset: TEM of Cu-doped ZnS NPs. Reproduced with permission (Ang et al. 2016). Copyright 2016, Wiley–VCH

et al. 2014; Zhou et al. 2014a). Pérez-Hernández et al. (2015) have been demonstrated (Fig. 2.1c, d). The interplay of NPs between imaging and therapy is depicted in a cartoon (Fig. 2.1e).

Some of the recent advancements of cancer nanomedicine include QD/Au NRs for imaging (Wu et al. 2015), NIR-responsive NPs for two or more combination therapy of cancer (Duan et al. 2017), and ultrasensitive fluorescence or magnetic based immunoassays for detection of folic acid (Li and Chen 2016) and C-reaction protein in clinical samples (Huang et al. 2018).

2.2   NPs as Drug Carriers

Although cancer nanomedicine has progressed rapidly over the years, it elicits challenges to combat toxicology issues (Chen et al. 2017; Shi et al. 2017; Youn and Bae 2018). Yet again, the targeted drug delivery is of paramount importance (Rosenblum et al. 2018).

2.2.1 Iron Oxide NPs

The intracellular uptake of superparamagnetic iron oxide NPs (SPIONs) is dictated by the ligands on their surface, thus facilitating drug delivery for cancer therapy (Huang et al. 2016; Luchini et al. 2017; Han et al. 2019). It has been found that phospholipids such as 1,2-dimyristoyl-sn-glycero-3-phosphocholine (DMPC) functionalized SPIONs are easily taken up by the PC-12 cells in larger amounts, compared to SPIONs modified with poly(ethylene glycol) (PEG) and polyethyleneimine (PEI) (Su et al. 2017). This study has concluded that DMPC-SPIONs can serve as potential drug carriers.

How do we suppress tumor growth? Several methods reported in the literature to address this issue. Once the designed NPs penetrate the tumor, they can suppress the breast tumor growth. Iron oxide NPs conjugated to p32 binding peptides with the sequence (AKRGARSTA) have been demonstrated to be most effective in treating breast cancer in mice via better tumor homing and penetration of the nanosystem (Sharma et al. 2017). In another interesting work, ultrafine iron oxide NPs (3.5 nm core size) have been shown to improve the delivery and intratumoral distribution and retention of NPs (Wang et al. 2017b). These small sized NPs in comparison with their large sized counterparts were found to be extravasated easily from the tumor vasculature, and diffused readily into the tumor tissue. In vivo MRI studies revealed that the ultrafine iron oxide NPs exhibited bright T1 contrast and dark T2 contrast in the tumor vasculature after 1 and 24 h of intravenous administration, respectively. The ultrafine iron oxide NPs with initial T1 contrast aggregated into larger clusters and exhibited T2 contrast.

2.2.2 Silica-Based Carriers

Functionalized mesoporous silica-based NPs and hollow materials emerged as transporters for targeted drug delivery (Tang et al. 2012; Zhang et al. 2012a; Li et al. 2017; Hai et al. 2018; Park and Ha 2018; Kesse et al. 2019). Multifunctional mesoporous silica NPs encompassing magnetic NPs and an anticancer drug (camptothecin) functionalized with TAT peptides and folic acid grafted chitosan have been designed to enable targeted drug delivery and MR imaging (Fig. 2.2a, b). This nanoassembly approach showed an enhanced anticancer effect, via the nucleus delivery of the DNA-toxin drug, inhibiting topoisomerase I and inducing cell apoptosis (Li et al. 2014).

Mesoporous silica NPs (core size of ~ 47–87 nm) modified with a crosslinked cationic polymer, polyethyleneimine–polyethylene-glycol copolymer, (PEI–PEG) was employed for efficient delivery of siRNA to HER2+ breast cancer (Ngamcherdtrakul et al. 2015). Polymer coated mesoporous silica NPs carried siRNA against the human epidermal growth factor (h-EGF) receptor type 2 (HER2) oncogene. This

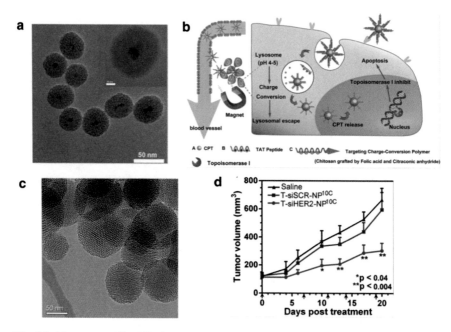

Fig. 2.2 Mesoporous silica NPs for cancer therapy. **a** TEM image of magnetic mesoporous silica NPs. **b** Cartoon depicting the magnetic assisted delivery and camptothecin (CPT) release into the nucleus by means of charge conversion polymer and lysosomal escape rendered by TAT peptides, and folic acid and citraconic anhydride grafted chitosan. Reproduced with permission (Li et al. 2014). Copyright 2014, Wiley–VCH. **c** TEM image of mesoporous silica NP (average particle size of 87 nm). **d** In vivo HER2 reduction and growth inhibition of orthotopic HCC1954 tumors. Tumor growth in mice bearing orthotopic HCC1954 tumor xenografts (n = 5/group) receiving the same treatments as (A) but multiple doses (days of injection are indicated by arrows). Reproduced with permission (Ngamcherdtrakul et al. 2015). Copyright 2014, Wiley–VCH

construct then coupled to anti-HER2 monoclonal antibody (trastuzumab). A scrambled siRNA (siSCR) and a siRNA against HER2 (siHER2) cross-linked with 10-kDa PEI, and coupled with trastuzumab, denoted as T–siSCR–NP 10C and T–siHER2–NP 10C, respectively (Fig. 2.2d). These trastuzumab-targeted siRNAs carrying mesoporous silica NPs could easily target the cancer cells that overexpress the HER2 protein.

2.2.3   Graphene and Graphene Oxide as Carriers

Graphene has superior qualities such as large surface area, low toxicity, good stability, and large drug loading capacity, and therefore proposed as a carrier for improved delivery of anticancer drug, doxorubicin (DOX) (Wang et al. 2014; Orecchioni et al. 2015). Graphene integrated with fluorescent QDs can act as reporters to monitor the

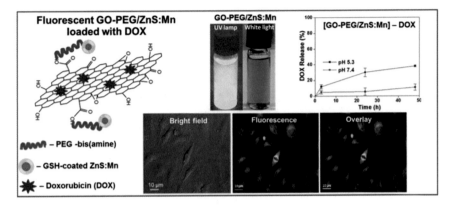

Fig. 2.3 Grafting of ZnS:Mn doped nanocrystal and anticancer drug onto graphene oxide for delivery and cell labelling. Reproduced with permission (Dinda et al. 2016). Copyright 2016, Wiley–VCH

drug delivery. To achieve specific targeting, a transferrin ligand (Trf) attached to the surface of the graphene sheet (Chen et al. 2013b).

Graphene oxide can act not only as a drug delivery vehicle but also as a substrate for the attachment of nanocrystals (Zhang et al. 2010; Zhou et al. 2014b). Very recently, we have reported the grafting of ZnS:Mn doped nanocrystal and anticancer drug onto graphene oxide for cancer cells labelling (Fig. 2.3) (Dinda et al. 2016). High drug entrapment efficiency, slow drug release, better cancer cell labelling and killing efficiency are the traits of this novel system. This demands further work for in-vivo animal imaging and targeted drug delivery. Besides, carbon nanodots have been exploited for the delivery of another anticancer drug, paclitaxel (PTX) (Gomez et al. 2018).

2.2.4 Peptide/Polymer Mediated Delivery

The efficient delivery of anticancer drugs (i.e., therapeutic efficacy) is an important goal in cancer nanomedicine. Towards this goal, an intra-nuclear drug delivery approach using dexamethasone-conjugated micelle has been developed (Wang et al. 2017a). Cell and tumor penetrating peptides have been designed for improved drug delivery (Ruoslahti 2017). Our group has developed QD-peptide bioconjugates for nuclear targeting of human mesenchymal stem cells (hMSCs) (Narayanan et al. 2013). To avoid the intracellular degradation of drugs in lysosomes and poor delivery to the nucleus, drugs can be encapsulated in copolymer micelles that are composed of N-(2-hydroxypropyl) methacrylamide (HPMA) (Zhou et al. 2017). The HA2 membrane fusion peptide grafted onto HPMA copolymers disrupted lysosome membranes. This work also illustrated that the micelles containing nucleus-targeting retinoic acid and the drug cargo (H1 peptide), efficiently evaded the lysosomes and targeted the nucleus of MCF-7 breast cancer cells, and further inhibited the tumor growth in mice.

Oligonucleotides possessing high immune-stimulatory activity such as cytosine-phosphate-guanine (CpG) conjugated to bifunctional NPs encompassing MnO_2–AgNCs–DOX conjugate can be used for enhanced cancer immunotherapy (Wang et al. 2017c). Here, MnO_2 sheets acted as unique support for the integration of chemotherapy drug DOX and the immunotherapeutic agent CpG-AgNCs. Furthermore, transport of NPs and drug mediated by DNA enabled endosomal escape and intracellular delivery (Muro 2014). A γ-glutamyl transpeptidase-responsive camptothecin–polymer conjugate for efficient suppression of solid tumors has been developed (Zhou et al. 2019). This conjugate also prolonged the survival of pancreatic tumour-bearing mice, compared to the chemotherapeutic drug gemcitabine.

2.2.5 Liposomes for Targeted Delivery

Liposomes are one of the emerging nanoscale drug delivery systems (Torchilin 2014; Zhao and Feng 2015; Vahed et al. 2017). They offer benefits such as low clearance rate, and improved targeting abilities. The potential target for siRNA-based cancer treatment is Bmi1 gene, which is overexpressed in various human tumors. Although siRNA-based therapy has potential merits in cancer treatment, it suffers from limited delivery targeting Bmi1 gene, low bioavailability and reduced efficacy. Liposome assisted co-delivery of folate-doxorubicin, and Bmi1 siRNA showed advantages in inhibiting the tumor growth via silencing the expression of Bmi1 gene and high targeting efficiency by folate receptor (Yang et al. 2014). Studies have shown that the anticancer drug, DOX did not respond to M109 lung tumors. However, a nanoassembly comprising of DOX with squalene, a natural lipid precursor for the biosynthesis of cholesterol, dramatically improved the anticancer efficacy, inhibiting the tumor by 90% (Maksimenko et al. 2014). This squalene-DOX nanomedicine reduced the cardiac toxicity induced by the drug and hence improved the therapeutic index of the drug.

PEGylated liposomal doxorubicin (PLD) is the first FDA-approved nanomedicine (Bobo et al. 2016) for cancer therapy of breast, ovarian, multiple myeloma, and Kaposi sarcoma (Gabizon et al. 2016). However, PLD has not replaced the conventional doxorubicin for the treatment of breast cancer, and its clinical application is not widespread. By exploiting the selective biodistribution and homing potential of PLD to tumors, and pharmaceutical nanotechnology with advanced theranostic platforms, will have great impact on future clinical applications.

2.3 Photothermal Therapy

Photothermal therapy (PTT) utilizes the NPs that can generate heat upon laser irradiation to induce apoptosis of cancer cells (Melamed et al. 2015). A wide variety of plasmonic materials such as Au, Ag, Cu, CuSe, graphene, carbon nanotubes have been exploited for photothermal cancer therapy. These plasmonic nanomaterials exhibit heat upon excitation to NIR light and less absorption by biological tissues.

Metallic nanostructures such as gold nanorods (Au-NRs) (Zhou et al. 2014a; Chen et al. 2013a), Au nanocages (Yavuz et al. 2009), Au–Cu alloy nanocrystals (He et al. 2014; Xia et al. 2017) exhibit absorption in the near-infrared (ca. 800–1200 nm) that allows for the excitation at this wavelength, thereby enabling photothermal effects, and photothermal therapy of cancer. Due to their large absorption cross sections in NIR region, Au-NRs (Zhang et al. 2012b), Au-prisms (Pérez-Hernández et al. 2014; Pelaz et al. 2012), Au-cages (Yavuz et al. 2009) and Au-nanostars (Yuan et al. 2012; Liu et al. 2015b; Liu et al. 2017b) have shown great potentials in photothermal therapy of cancer. All these Au-NRs, Au-prisms and Au-cages can act as nanoheaters under NIR illumination, exhibiting promising photothermal application in nanomedicine. Besides Au nanostructures, bismuth sulfide nanorods (Liu et al. 2015a), and Te-NRs (decorated by polysaccharide–protein complex) (Huang et al. 2017) have been used for combined chemo and photothermal therapy.

Recent work has demonstrated the photothermal therapy of PTW-Te-NRs (PTW: extracted from *Pleurotus tuber-reguim*) (Huang et al. 2017). The effects of Te-NPs and PTW-Te-NRs on HepG2 cells after laser irradiation for different time intervals captured by IR camera indicated a sharp rise in temperature with the maximum of 68.2 °C (Fig. 2.4a, b). The fact that the reduction in cell number and rounded cell shape for both Te-NPs and PTW-TeNRs with laser treated cells clearly indicated the apoptosis of cancer cells (Fig. 2.4c). The fluorescence staining (live and dead cell assays) with calcein acetoxymethyl (green) and propidium iodide (red) further supported the above observation (Fig. 2.4d). This study demonstrated the potential of these Te-NRs for chemo-photothermal combination therapy of cancer (Fig. 2.4e).

Other potential NP systems are plasmonic bimetallic Au–Cu nanocrystals for combined chemo and PTT (He et al. 2014; Nanda et al. 2019), core–shell Au–Se nanocomposites for radio- and chemo-therapy (Chang et al. 2017), and targeted combination therapy of cancer using NIR dye (ICG: indocyanine green)—bovine serum albumin nanocomplex as photothermal agent and DOX as chemotherapy drug, co-encapsulated into engineered red blood cells (RBCs) carriers (Sun et al. 2015). In the later system, RBCs heated up due to NIR irradiation, resulting in a burst drug release (ca. 80%) within 5 min.

In conclusion, cancer nanomedicine shows promises in early diagnostics and preoperative therapeutics, neoadjuvant radiotherapy, chemotherapy, phototherapy, and immunotherapy (Qu et al. 2023).

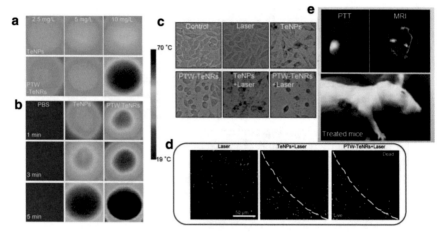

Fig. 2.4 Tellurium nanorods (Te-NRs) for PTT. **a** IR camera captured thermal images of HepG2 cells incubated with either Te-NPs or PTW-Te-NRs with specified concentrations upon NIR light irradiation for 5 min. **b** Temperature profiles of NPs and NRs (20 mg L^{-1}) exposed HepG2 cells recorded at 1, 3 and 5 min of irradiation (808 nm, 3 W cm^{-2}). **c** Bright-field images of HepG2 cells with control groups (PBS, laser only, free TeNPs, free PTW − TeNRs) and laser irradiated cells (TeNPs + Laser, and PTW − TeNRs + Laser). **d** Fluorescence images of HepG2 cells treated with TeNPs or PTW-TeNRs, and stained with calcein AM (green) and PI (red). **e** In vivo photothermal therapy (PTT) and MRI of TeNRs (PTW − TeNRs + Laser) treated mice after 21 days of observation. Reproduced with permission (Huang et al. 2017). Copyright 2017, Wiley–VCH

References

Ali MR, Wu Y, Ghosh D, Do BH, Chen K, Dawson MR, Fang N, Sulchek TA, El-Sayed MA (2017) Nuclear membrane-targeted gold nanoparticles inhibit cancer cell migration and invasion. ACS Nano 11(4):3716–3726

Ang H, Bosman M, Thamankar R, Zulkifli MFB, Yen SK, Hariharan A, Sudhaharan T, Selvan ST (2016) Highly luminescent heterostructured copper-doped zinc sulfide nanocrystals for application in cancer cell labeling. ChemPhysChem 17(16):2489–2495

Bobo D, Robinson KJ, Islam J, Thurecht KJ, Corrie SR (2016) Nanoparticle-based medicines: a review of FDA-approved materials and clinical trials to date. Pharm Res 33(10):2373–2387

Brunden KR, Lee VM, Smith AB III, Trojanowski JQ, Ballatore C (2017) Altered microtubule dynamics in neurodegenerative disease: therapeutic potential of microtubule-stabilizing drugs. Neurobiol Dis 105:328–335

Chang Y, He L, Li Z, Zeng L, Song Z, Li P, Chan L, You Y, Yu X-F, Chu PK (2017) Designing core–shell gold and selenium nanocomposites for cancer radiochemotherapy. ACS Nano 11(5):4848–4858

Chen H, Shao L, Li Q, Wang J (2013a) Gold nanorods and their plasmonic properties. Chem Soc Rev 42(7):2679–2724

Chen M-L, He Y-J, Chen X-W, Wang J-H (2013b) Quantum-dot-conjugated graphene as a probe for simultaneous cancer-targeted fluorescent imaging, tracking, and monitoring drug delivery. Bioconjug Chem 24(3):387–397

Chen F, Goel S, Hernandez R, Graves SA, Shi S, Nickles RJ, Cai W (2016) Dynamic positron emission tomography imaging of renal clearable gold nanoparticles. Small 12(20):2775–2782

Chen H, Zhang W, Zhu G, Xie J, Chen X (2017) Rethinking cancer nanotheranostics. Nat Rev Mater 2(7):17024

Cheng K, Kothapalli S-R, Liu H, Koh AL, Jokerst JV, Jiang H, Yang M, Li J, Levi J, Wu JC (2014) Construction and validation of nano gold tripods for molecular imaging of living subjects. J Am Chem Soc 136(9):3560–3571

Dearling JL, Packard AB (2017) Molecular imaging in nanomedicine—a developmental tool and a clinical necessity. J Control Release 261:23–30

Dinda S, Kakran M, Zeng J, Sudhaharan T, Ahmed S, Das D, Selvan ST (2016) Grafting of ZnS: Mn-doped nanocrystals and an anticancer drug onto graphene oxide for delivery and cell labeling. ChemPlusChem 81(1):100–107

Ding C, Cao X, Zhang C, He T, Hua N, Xian Y (2017) Rare earth ions enhanced near infrared fluorescence of Ag_2S quantum dots for the detection of fluoride ions in living cells. Nanoscale 9(37):14031–14038

Duan S, Yang Y, Zhang C, Zhao N, Xu FJ (2017) NIR-responsive polycationic gatekeeper-cloaked hetero-nanoparticles for multimodal imaging-guided triple-combination therapy of cancer. Small 13(9):1603133

Gabizon AA, Patil Y, La-Beck NM (2016) New insights and evolving role of pegylated liposomal doxorubicin in cancer therapy. Drug Resist Updates 29:90–106

Ganguly S, Jacob SK (2017) Therapeutic outlook of pyrazole analogs: a mini review. Mini Rev Med Chem 17(11):959–983

Gomez IJ, Arnaiz B, Cacioppo M, Arcudi F, Prato M (2018) Nitrogen-doped carbon nanodots for bioimaging and delivery of paclitaxel. J Mater Chem B 6(35):5540–5548

Gong J, Chen M, Zheng Y, Wang S, Wang Y (2012) Polymeric micelles drug delivery system in oncology. J Control Release 159(3):312–323

Gu Y-P, Cui R, Zhang Z-L, Xie Z-X, Pang D-W (2011) Ultrasmall near-infrared Ag_2Se quantum dots with tunable fluorescence for in vivo imaging. J Am Chem Soc 134(1):79–82

Hai L, Jia X, He D, Zhang A, Wang T, Cheng H, He X, Wang K (2018) DNA-functionalized hollow mesoporous silica nanoparticles with dual cargo loading for near-infrared-responsive synergistic chemo-photothermal treatment of cancer cells. ACS Appl Nano Mater 1(7):3486–3497

Han G, Zhang B, Zhang H, Han D, Tan J, Yang B (2019) The synthesis and characterization of glutathione-modified superparamagnetic iron oxide nanoparticles and their distribution in rat brains after injection in substantia nigra. J Mater Sci—Mater Med 30(1):5

He R, Wang Y-C, Wang X, Wang Z, Liu G, Zhou W, Wen L, Li Q, Wang X, Chen X (2014) Facile synthesis of pentacle gold–copper alloy nanocrystals and their plasmonic and catalytic properties. Nat Commun 5:4327

Hong G, Robinson JT, Zhang Y, Diao S, Antaris AL, Wang Q, Dai H (2012) In vivo fluorescence imaging with Ag_2S quantum dots in the second near-infrared region. Angew Chem Int Ed 51(39):9818–9821

Hong G, Antaris AL, Dai H (2017) Near-infrared fluorophores for biomedical imaging. Nat Biomed Eng 1(1):0010

Huang J, Li Y, Orza A, Lu Q, Guo P, Wang L, Yang L, Mao H (2016) Magnetic nanoparticle facilitated drug delivery for cancer therapy with targeted and image-guided approaches. Adv Func Mater 26(22):3818–3836

Huang W, Huang Y, You Y, Nie T, Chen T (2017) High-yield synthesis of multifunctional tellurium nanorods to achieve simultaneous chemo-photothermal combination cancer therapy. Adv Func Mater 27(33):1701388

Huang L, Liao T, Wang J, Ao L, Su W, Hu J (2018) Brilliant pitaya-type silica colloids with central-radial and high-density quantum dots incorporation for ultrasensitive fluorescence immunoassays. Adv Func Mater 28(4):1705380

Hwang H, Jeong H-S, Oh P-S, Kim M, Lee T-K, Kwon J, Kim H-S, Lim ST, Sohn M-H, Jeong H-J (2016) PEGylated nanoliposomes encapsulating angiogenic peptides improve perfusion defects: radionuclide imaging-based study. Nucl Med Biol 43(9):552–558

Kataoka K, Harada A, Nagasaki Y (2012) Block copolymer micelles for drug delivery: design, characterization and biological significance. Adv Drug Deliv Rev 64:37–48

Kesse S, Boakye-Yiadom KO, Ochete BO, Opoku-Damoah Y, Akhtar F, Filli MS, Asim Farooq M, Aquib M, Mily M, Joelle B (2019) Mesoporous silica nanomaterials: versatile nanocarriers for cancer theranostics and drug and gene delivery. Pharmaceutics 11(2):77

Khandelwal P, Poddar P (2017) Fluorescent metal quantum clusters: an updated overview of the synthesis, properties, and biological applications. J Mater Chem B 5(46):9055–9084

Komatsu N (2023) Poly (glycerol)-based biomedical nanodevices constructed by functional programming on inorganic nanoparticles for cancer nanomedicine. Acc Chem Res 1388–1392

Kuppast B, Fahmy H (2016) Thiazolo [4, 5-d] pyrimidines as a privileged scaffold in drug discovery. Eur J Med Chem 113:198–213

Li X, Chen L (2016) Fluorescence probe based on an amino-functionalized fluorescent magnetic nanocomposite for detection of folic acid in serum. ACS Appl Mater Interfaces 8(46):31832–31840

Li Z, Dong K, Huang S, Ju E, Liu Z, Yin M, Ren J, Qu X (2014) A smart nanoassembly for multistage targeted drug delivery and magnetic resonance imaging. Adv Func Mater 24(23):3612–3620

Li Y, Li N, Pan W, Yu Z, Yang L, Tang B (2017) Hollow mesoporous silica nanoparticles with tunable structures for controlled drug delivery. ACS Appl Mater Interfaces 9(3):2123–2129

Liu J, Yu M, Zhou C, Yang S, Ning X, Zheng J (2013) Passive tumor targeting of renal-clearable luminescent gold nanoparticles: long tumor retention and fast normal tissue clearance. J Am Chem Soc 135(13):4978–4981

Liu J, Zheng X, Yan L, Zhou L, Tian G, Yin W, Wang L, Liu Y, Hu Z, Gu Z (2015a) Bismuth sulfide nanorods as a precision nanomedicine for in vivo multimodal imaging-guided photothermal therapy of tumor. ACS Nano 9(1):696–707

Liu Y, Ashton JR, Moding EJ, Yuan H, Register JK, Fales AM, Choi J, Whitley MJ, Zhao X, Qi Y (2015b) A plasmonic gold nanostar theranostic probe for in vivo tumor imaging and photothermal therapy. Theranostics 5(9):946

Liu T-M, Conde J, Lipiński T, Bednarkiewicz A, Huang C-C (2017a) Smart NIR linear and nonlinear optical nanomaterials for cancer theranostics: prospects in photomedicine. Prog Mater Sci 88:89–135

Liu Y, Maccarini P, Palmer GM, Etienne W, Zhao Y, Lee C-T, Ma X, Inman BA, Vo-Dinh T (2017b) Synergistic immuno photothermal nanotherapy (SYMPHONY) for the treatment of unresectable and metastatic cancers. Sci Rep 7(1):8606

Luchini A, Gerelli Y, Fragneto G, Nylander T, Pálsson GK, Appavou M-S, Paduano L (2017) Neutron reflectometry reveals the interaction between functionalized SPIONs and the surface of lipid bilayers. Colloids Surf, B 151:76–87

Maksimenko A, Dosio F, Mougin J, Ferrero A, Wack S, Reddy LH, Weyn A-A, Lepeltier E, Bourgaux C, Stella B (2014) A unique squalenoylated and nonpegylated doxorubicin nanomedicine with systemic long-circulating properties and anticancer activity. Proc Natl Acad Sci 111(2):E217–E226

Malinge J, Géraudie B, Savel P, Nataf V, Prignon A, Provost C, Zhang Y, Ou P, Kerrou K, Talbot J-N (2017) Liposomes for PET and MR imaging and for dual targeting (magnetic field/glucose moiety): synthesis, properties, and in vivo studies. Mol Pharm 14(2):406–414

Man F, Gawne PJ, de Rosales RT (2019) Nuclear imaging of liposomal drug delivery systems: a critical review of radiolabelling methods and applications in nanomedicine. Adv Drug Deliv Rev

Melamed JR, Edelstein RS, Day ES (2015) Elucidating the fundamental mechanisms of cell death triggered by photothermal therapy. ACS Nano 9(1):6–11

Meng H, Mai WX, Zhang H, Xue M, Xia T, Lin S, Wang X, Zhao Y, Ji Z, Zink JI (2013) Codelivery of an optimal drug/siRNA combination using mesoporous silica nanoparticles to overcome drug resistance in breast cancer in vitro and in vivo. ACS Nano 7(2):994–1005

Miller MA, Chandra R, Cuccarese MF, Pfirschke C, Engblom C, Stapleton S, Adhikary U, Kohler RH, Mohan JF, Pittet MJ (2017) Radiation therapy primes tumors for nanotherapeutic delivery via macrophage-mediated vascular bursts. Sci Transl Med 9(392):eaal0225

Muro S (2014) A DNA device that mediates selective endosomal escape and intracellular delivery of drugs and biologicals. Adv Func Mater 24(19):2899–2906

Nanda SS, Hembram K, Lee J-K, Kim K, Selvan ST, Yi DK (2019) Experimental and theoretical structural characterization of Cu–Au tripods for photothermal anticancer therapy. ACS Appl Nano Materr 2(6):3735–3742

Narayanan K, Yen SK, Dou Q, Padmanabhan P, Sudhaharan T, Ahmed S, Ying JY, Selvan ST (2013) Mimicking cellular transport mechanism in stem cells through endosomal escape of new peptide-coated quantum dots. Sci Rep 3:2184

Nasongkla N, Bey E, Ren J, Ai H, Khemtong C, Guthi JS, Chin S-F, Sherry AD, Boothman DA, Gao J (2006) Multifunctional polymeric micelles as cancer-targeted, MRI-ultrasensitive drug delivery systems. Nano Lett 6(11):2427–2430

Ngamcherdtrakul W, Morry J, Gu S, Castro DJ, Goodyear SM, Sangvanich T, Reda MM, Lee R, Mihelic SA, Beckman BL (2015) Cationic polymer modified mesoporous silica nanoparticles for targeted siRNA delivery to HER2+ breast cancer. Adv Func Mater 25(18):2646–2659

Orecchioni M, Cabizza R, Bianco A, Delogu LG (2015) Graphene as cancer theranostic tool: progress and future challenges. Theranostics 5(7):710

Park SS, Ha CS (2018) Hollow mesoporous functional hybrid materials: fascinating platforms for advanced applications. Adv Func Mater 28(27):1703814

Pelaz B, Grazu V, Ibarra A, Magen C, del Pino P, de la Fuente JM (2012) Tailoring the synthesis and heating ability of gold nanoprisms for bioapplications. Langmuir 28(24):8965–8970

Pérez-Hernández M, del Pino P, Mitchell SG, Moros M, Stepien G, Pelaz B, Parak WJ, Gálvez EM, Pardo J, de la Fuente JM (2014) Dissecting the molecular mechanism of apoptosis during photothermal therapy using gold nanoprisms. ACS nano 9(1):52–61

Pérez-Hernández M, del Pino P, Mitchell SG, Moros M, Stepien G, Pelaz B, Parak WJ, Gálvez EM, Pardo J, de la Fuente JM (2015) Dissecting the molecular mechanism of apoptosis during photothermal therapy using gold nanoprisms. ACS Nano 9(1):52–61

Pratt EC, Shaffer TM, Grimm J (2016) Nanoparticles and radiotracers: advances toward radio-nanomedicine. Wiley Interdisc Rev: Nanomed Nanobiotechnol 8(6):872–890

Pyo K, Thanthirige VD, Kwak K, Pandurangan P, Ramakrishna G, Lee D (2015) Ultrabright luminescence from gold nanoclusters: rigidifying the Au (I)–thiolate shell. J Am Chem Soc 137(25):8244–8250

Pyo K, Ly NH, Yoon SY, Shen Y, Choi SY, Lee SY, Joo SW, Lee D (2017) Highly luminescent folate-functionalized Au_{22} nanoclusters for bioimaging. Adv Healthcare Mater 6(16):1700203

Qu X, Hu Y, Wang H, Song H, Young M, Xu F, Liu Y, Cheng G (2019) Biomimetic dextran-peptide vectors for efficient and safe siRNA delivery. ACS Appl Bio Mater 2(4):1456–1463

Qu X, Zhou D, Lu J, Qin D, Zhou J, Liu H-J (2023) Cancer nanomedicine in preoperative thera-peutics: nanotechnology-enabled neoadjuvant chemotherapy, radiotherapy, immunotherapy, and phototherapy. Bioact Mater 24:136–152

Rosenblum D, Joshi N, Tao W, Karp JM, Peer D (2018) Progress and challenges towards targeted delivery of cancer therapeutics. Nat Commun 9(1):1410

Ruiz-Ceja KA, Chirino YI (2017) Current FDA-approved treatments for non-small cell lung cancer and potential biomarkers for its detection. Biomed Pharmacother 90:24–37

Ruoslahti E (2017) Tumor penetrating peptides for improved drug delivery. Adv Drug Deliv Rev 110:3–12

Shang L, Stockmar F, Azadfar N, Nienhaus GU (2013) Intracellular thermometry by using fluorescent gold nanoclusters. Angew Chem Int Ed 52(42):11154–11157

Sharma S, Kotamraju VR, Mölder T, Tobi A, Teesalu T, Ruoslahti E (2017) Tumor-penetrating nanosystem strongly suppresses breast tumor growth. Nano Lett 17(3):1356–1364

Shi J, Kantoff PW, Wooster R, Farokhzad OC (2017) Cancer nanomedicine: progress, challenges and opportunities. Nat Rev Cancer 17(1):20

Su L, Zhang B, Huang Y, Fan Z, Zhao Y (2017) Enhanced cellular uptake of iron oxide nanoparticles modified with 1, 2-dimyristoyl-sn-glycero-3-phosphocholine. RSC Adv 7(60):38001–38007

Sun X, Wang C, Gao M, Hu A, Liu Z (2015) Remotely controlled red blood cell carriers for cancer targeting and near-infrared light-triggered drug release in combined photothermal-chemotherapy. Adv Func Mater 25(16):2386–2394

Sunoqrot S, Hamed R, Abdel-Halim H, Tarawneh O (2017) Synergistic interplay of medicinal chemistry and formulation strategies in nanotechnology-from drug discovery to nanocarrier design and development. Curr Top Med Chem 17(13):1451–1468

Tang F, Li L, Chen D (2012) Mesoporous silica nanoparticles: synthesis, biocompatibility and drug delivery. Adv Mater 24(12):1504–1534

Tian R, Shen Z, Zhou Z, Munasinghe J, Zhang X, Jacobson O, Zhang M, Niu G, Pang DW, Cui R (2019) Ultrasmall quantum dots with broad-spectrum metal doping ability for trimodal molecular imaging. Adv Funct Mater 1901671

Torchilin VP (2014) Multifunctional, stimuli-sensitive nanoparticulate systems for drug delivery. Nat Rev Drug Discov 13(11):813

Vahed SZ, Salehi R, Davaran S, Sharifi S (2017) Liposome-based drug co-delivery systems in cancer cells. Mater Sci Eng, C 71:1327–1341

Wang X, Sun X, Lao J, He H, Cheng T, Wang M, Wang S, Huang F (2014) Multifunctional graphene quantum dots for simultaneous targeted cellular imaging and drug delivery. Colloids Surf, B 122:638–644

Wang H, Li Y, Bai H, Shen J, Chen X, Ping Y, Tang G (2017a) A cooperative dimensional strategy for enhanced nucleus-targeted delivery of anticancer drugs. Adv Func Mater 27(24):1700339

Wang L, Huang J, Chen H, Wu H, Xu Y, Li Y, Yi H, Wang YA, Yang L, Mao H (2017b) Exerting enhanced permeability and retention effect driven delivery by ultrafine iron oxide nanoparticles with T 1–T 2 switchable magnetic resonance imaging contrast. ACS Nano 11(5):4582–4592

Wang Z, Zhang Y, Liu Z, Dong K, Liu C, Ran X, Pu F, Ju E, Ren J, Qu X (2017c) A bifunctional nanomodulator for boosting CpG-mediated cancer immunotherapy. Nanoscale 9(37):14236–14247

Wu Q, Chen L, Huang L, Wang J, Liu J, Hu C, Han H (2015) Quantum dots decorated gold nanorod as fluorescent-plasmonic dual-modal contrasts agent for cancer imaging. Biosens Bioelectron 74:16–23

Xia Y, Gilroy KD, Peng HC, Xia X (2017) Seed-mediated growth of colloidal metal nanocrystals. Angew Chem Int Ed 56(1):60–95

Xing H, Bu W, Zhang S, Zheng X, Li M, Chen F, He Q, Zhou L, Peng W, Hua Y (2012) Multifunctional nanoprobes for upconversion fluorescence, MR and CT trimodal imaging. Biomaterials 33(4):1079–1089

Yahia-Ammar A, Sierra D, Mérola F, Hildebrandt N, Le Guével X (2016) Self-assembled gold nanoclusters for bright fluorescence imaging and enhanced drug delivery. ACS Nano 10(2):2591–2599

Yang T, Li B, Qi S, Liu Y, Gai Y, Ye P, Yang G, Zhang W, Zhang P, He X (2014) Co-delivery of doxorubicin and Bmi1 siRNA by folate receptor targeted liposomes exhibits enhanced anti-tumor effects in vitro and in vivo. Theranostics 4(11):1096

Yang F, Skripka A, Tabatabaei MS, Hong SH, Ren F, Benayas A, Oh JK, Martel S, Liu X, Vetrone F (2019) Multifunctional self-assembled supernanoparticles for deep-tissue bimodal imaging and amplified dual-mode heating treatment. ACS Nano 13(1):408–420

Yavuz MS, Cheng Y, Chen J, Cobley CM, Zhang Q, Rycenga M, Xie J, Kim C, Song KH, Schwartz AG (2009) Gold nanocages covered by smart polymers for controlled release with near-infrared light. Nat Mater 8(12):935

Yen SK, Janczewski D, Lakshmi JL, Dolmanan SB, Tripathy S, Ho VH, Vijayaragavan V, Hariharan A, Padmanabhan P, Bhakoo KK (2013) Design and synthesis of polymer-functionalized NIR fluorescent dyes–magnetic nanoparticles for bioimaging. ACS Nano 7(8):6796–6805

Youn YS, Bae YH (2018) Perspectives on the past, present, and future of cancer nanomedicine. Adv Drug Deliv Rev 130:3–11

Yu Y, Luo Z, Chevrier DM, Leong DT, Zhang P, Jiang D-E, Xie J (2014) Identification of a highly luminescent $Au_{22}(SG)_{18}$ nanocluster. J Am Chem Soc 136(4):1246–1249

Yuan H, Fales AM, Vo-Dinh T (2012) TAT peptide-functionalized gold nanostars: enhanced intracellular delivery and efficient NIR photothermal therapy using ultralow irradiance. J Am Chem Soc 134(28):11358–11361

Zhang L, Xia J, Zhao Q, Liu L, Zhang Z (2010) Functional graphene oxide as a nanocarrier for controlled loading and targeted delivery of mixed anticancer drugs. Small 6(4):537–544

Zhang L, Lu Z, Zhao Q, Huang J, Shen H, Zhang Z (2011) Enhanced chemotherapy efficacy by sequential delivery of siRNA and anticancer drugs using PEI-grafted graphene oxide. Small 7(4):460–464

Zhang Q, Liu F, Nguyen KT, Ma X, Wang X, Xing B, Zhao Y (2012a) Multifunctional mesoporous silica nanoparticles for cancer-targeted and controlled drug delivery. Adv Func Mater 22(24):5144–5156

Zhang Z, Wang L, Wang J, Jiang X, Li X, Hu Z, Ji Y, Wu X, Chen C (2012b) Mesoporous silica-coated gold nanorods as a light-mediated multifunctional theranostic platform for cancer treatment. Adv Mater 24(11):1418–1423

Zhang C, Li C, Liu Y, Zhang J, Bao C, Liang S, Wang Q, Yang Y, Fu H, Wang K (2015) Gold nanoclusters-based nanoprobes for simultaneous fluorescence imaging and targeted photodynamic therapy with superior penetration and retention behavior in tumors. Adv Func Mater 25(8):1314–1325

Zhao J, Feng S-S (2015) Nanocarriers for delivery of siRNA and co-delivery of siRNA and other therapeutic agents. Nanomedicine 10(14):2199–2228

Zhao Y, Mu X, Du G (2016) Microtubule-stabilizing agents: new drug discovery and cancer therapy. Pharmacol Ther 162:134–143

Zhou T, Yu M, Zhang B, Wang L, Wu X, Zhou H, Du Y, Hao J, Tu Y, Chen C (2014a) Inhibition of cancer cell migration by gold nanorods: molecular mechanisms and implications for cancer therapy. Adv Func Mater 24(44):6922–6932

Zhou T, Zhou X, Xing D (2014b) Controlled release of doxorubicin from graphene oxide based charge-reversal nanocarrier. Biomaterials 35(13):4185–4194

Zhou Z, Liu Y, Wu L, Li L, Huang Y (2017) Enhanced nuclear delivery of anti-cancer drugs using micelles containing releasable membrane fusion peptide and nuclear-targeting retinoic acid. J Mater Chem B 5(34):7175–7185

Zhou Q, Shao S, Wang J, Xu C, Xiang J, Piao Y, Zhou Z, Yu Q, Tang J, Liu X (2019) Enzyme-activatable polymer–drug conjugate augments tumour penetration and treatment efficacy. Nat Nanotechnol 14(8):799–809

Chapter 3
Nanomedicine for Neurodegenerative Diseases

Abstract This chapter deals with the design of organic NPs (e.g., curcumin, green tea polyphenol—EGCG), and inorganic NPs (e.g., Au, ZnO, CeO_2) in decreasing or inhibiting the amyloid aggregation and tau hyperphosphorylation associated with the Alzheimer's disease (AD). Different NP-based drug delivery approaches (e.g., apolipoprotein, peptides, dendrimers) to the delivery of CNS drugs across the blood–brain barrier (BBB) are discussed.

Keywords Ceria nanoparticles · ROS scavengers · Nanomedicine · Targeted delivery · Diagnostics · Neurodegenerative diseases · Alzheimer's disease · Parkinson's disease · Biosensors

The most common neurodegenerative diseases are Alzheimer's disease (AD) and Parkinson's disease (PD). Other common neurological disorders are acute spinal cord injury, epilepsy and seizures, migraines, multiple sclerosis, brain tumors, and stroke. Some of these degenerative nerve diseases are genetic, and they affect the activities of our body such as balance, movement, talking, breathing, and heart function. The main causes of the above neurodegenerative diseases are mostly unknown; however, it could originate from a medical condition (e.g., stroke, tumor or alcoholism), or toxins, and viruses. Importantly, although brain, spinal cord, and nerves (central nervous system, CNS) are safeguarded by the meninges, serious bacterial, viral, or fungal infection in the brain can cause life-threatening diseases such as meningitis.

It is highly indispensable that the CNS drugs should pass through the blood–brain barrier (BBB). Engineered nanostructures and nanomaterials hold great promise for both diagnosis and therapeutic applications. They can also be useful for combating microbial drug resistance, due to their high surface area and innate antibacterial activity. Several nanoparticle-based approaches have been delineated to enhance the CNS delivery of drugs across BBB (Tamil Selvan et al. 2020).

In the United States itself, over 6.2 million people suffered from AD in 2022, according to a report from the Alzheimer's Disease Association. Neurodegenerative AD correlates closely to the aggregation of amyloid beta (Aβ) proteins and hyper-phosphorylated tau protein, neurofibrillary tangles (NFTs). However, the

S. Tamil Selvan, *Nanomedicine*, Nanotheranostics,
https://doi.org/10.1007/978-981-99-2139-3_3

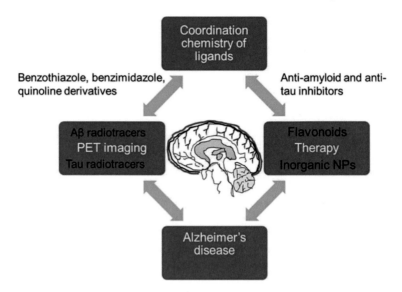

Fig. 3.1 The interplay of ligands in the design of tracers for imaging, and organic/inorganic NPs for therapeutic applications of AD. Reproduced with permission (Tamil Selvan et al. 2021) Copyright 2021, Elsevier

amyloid aggregation is the early event in AD, and its progression (memory cognitive impairment, MCI) is associated to the tau NFTs.

Importantly, nanomaterials pave their way to combat the protein amyloid aggregation (Wang et al. 2017). For instance, maghemite iron oxide NPs coated with dextran polymer inhibited the amyloid fibrillogenesis of human insulin (Lu et al. 2018). Nanomedicine for neurodegenerative diseases has been emerging with an aim to ameliorate the neuroprotection and combat the difficulties associated with the passage of neuro-protecting agents through the blood–brain-barrier (BBB). Recently, different nano-based systems for enhancing the neuroprotective efficacy have been developed. In a recent review, we have discussed the effects of coordination ligands on the surface of organic NPs such as flavonoids (e.g., curcumin, green tea polyphenol—EGCG,), and inorganic NPs (e.g., Au, ZnO, CeO_2) in decreasing or inhibiting the amyloid aggregation and tau hyperphosphorylation (Fig. 3.1) (Tamil Selvan et al. 2021).

3.1 Design of NPs for Inhibiting Protein Aggregation

The new term "nano-neuroscience" is an emerging paradigm, which bridges two burgeoning fields of nano- and neuro-science (Kumar, Tan et al. 2017). The probe development with the advent of nanotechnology opens up new avenues for diagnosing and treating neurodegenerative diseases. Nanostructures with binding ability to aggregated Aβ proteins and permeability to the BBB are useful nanomedicine platforms (Goldsmith et al. 2014; Huang et al. 2015; Aparicio-Blanco et al. 2016).

3.1.1 Au NPs

Several NP-based strategies have been designed to prevent the tau protein hyper-phosphorylation, so that apoptosis and neurodegeneration can be inhibited. The anthocyanin-loaded PEG-Au NPs prevented the hyper-phosphorylation of tau protein, thereby enhancing the neuroprotection in an $Aβ_{1-42}$ mouse model of Alzheimer's disease (AD) (Ali et al. 2017). This is considered as one of the promising nanomedicine strategies in preventing neurodegenerative diseases. Another study suggested that anthocyanins act as effective anti-oxidant neuroprotective agent in combating oxidative stress, thus improving memory impairments (Ali et al. 2018). Such Au NPs are good candidate systems for traversing the BBB to probe the brain-tumor surgery (Gao et al. 2017).

Conjugation of two peptide inhibitors (VVIA and LPFFD) onto Au-NPs reduced the cytotoxicity caused by Aβ aggregation (Xiong et al. 2017). Jana and co-workers have found that water-soluble curcumin-functionalized Au-NPs (Au-curcumin) could inhibit amyloid fibrillation and disintegrate or dissolve amyloid fibrils. This is a promising therapeutic approach to the neurodegenerative diseases (Palmal et al. 2014). In vitro inhibition of Aβ peptide aggregation into fibrils by noble metal Au and Pt NPs indicated clearly their binding affinity to amyloids (Streich et al. 2016).

3.1.2 Ceria NPs

Cerium oxide or ceria (CeO_2) NPs emerge as an effective antioxidant for various neurodegenerative diseases including Alzheimer's and Parkinson's diseases, ischemic stroke, and multiple sclerosis (Naz et al. 2017). Surface-functionalized ceria NPs have mitigated the mitochondrial oxidative stress and suppressed tau hyper-phosphorylation (Chen et al. 2018). The nanocomposite was fabricated by assembling ceria (CeNC) and iron oxide nanocrystals (IONC) onto mesoporous silica NPs. The addition of preformed NOTA-T807 (NOTA: 1,4,7-triazacyclononane-1,4,7-triacetic acid; T807: tau PET tracer) and methylene blue (MB) completed the formation of multifunctional (CeNC/IONC/MSN-T807-MB) nanocomposite. The nanocomposite labelled with ^{68}Ga and NOTA-T807 allowed for active targeting and imaging of hyper-phosphorylated tau by PET and MRI, and combinational therapy of ROS scavenging and MB release (Fig. 3.2). This work sheds light in preventing mitochondrial oxidative stress induced hyper-phosphorylation of tau (Chen et al. 2018).

In another interesting work, ceria NPs were employed for scavenging intra- and extra-cellular and mitochondrial ROS in PD model mice (Fig. 3.3) (Kwon et al. 2018).

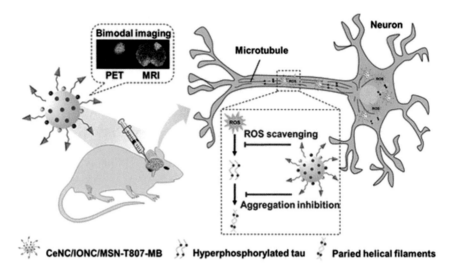

Fig. 3.2 Ceria NPs for the scavenging of ROS in mitigating oxidative stress in AD. Design of multifunctional NPs (CeNC/IONC/MSN-T807-MB) for targeting hyper-phosphorylated tau and combinational therapy of ROS scavenging and methylene blue release. Reproduced with permission (Chen et al. 2018). Copyright 2018, ACS

The ability of ceria NPs in scavenging ROS redox reactions (switching between Ce^{3+} and Ce^{4+} ions) is based on the enzymatic (SOD and catalase) mimetic activity (Fig. 3.3A). However, the cell uptake is dependent on the particle size, surface coating and charge. Ceria NPs possessing negative charge labelled SH-SY5Y cells cytoplasm, but those coated with triphenylphosphonium (TPP) and PEG labelled the mitochondria, as inferred from confocal images (Fig. 3.3B).

Immunohistochemistry (IHC) and fluorescence of the brain sections visualized the distribution of tyrosine hydroxylase (TH) and showed higher levels in in the striatal regions of the PD for ceria NP-treated groups (Fig. 3.3C). This work demonstrated the usefulness of ceria NPs in inhibiting the microglial activation and lipid peroxidation via the scavenging of intracellular or mitochondrial ROS, and in protecting the tyrosine hydroxylase in the striatal of regions of the PD mice.

Importantly, ceria NPs showed excellent antioxidant properties to combat reactive oxygen species (ROS) induced oxidative stress by targeting mitochondria in AD (Kwon et al. 2016), protection against ischemic stroke (Kim et al. 2012), and for selective scavenging of mitochondrial ROS in Parkinson's disease (Kwon et al. 2018). Furthermore, bimodal ceria/magnetite (Kim et al. 2019), and Ce/MnMoS$_4$ (Guan et al. 2016) core–shell NPs have been developed for addressing multiple facets of AD. Conversely, ceria NPs served as neuroprotection agents (Rzigalinski et al. 2017), and used to ameliorate the neurochemical impairments (induced by hydroxydopamine) in PD rats (Hegazy et al. 2017).

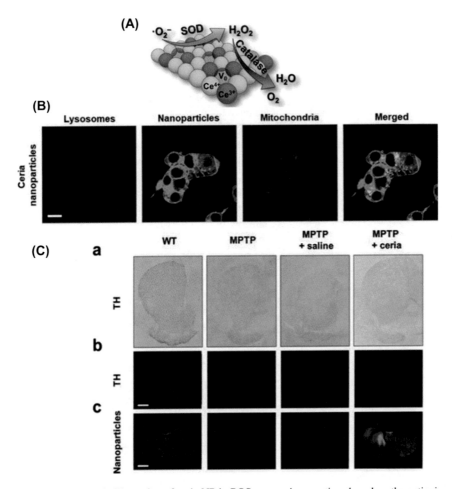

Fig. 3.3 A Schematic illustration of ceria NP in ROS scavenging reactions based on the antioxidant enzymes—super oxide dismutase (SOD) and catalase. **B** Confocal images of SH-SY5Y cells labelled with ceria NPs (blue: lysosomes; red: mitochondria, green: FITC-conjugated NPs). Scale bars are 10 μm. **C** Tyrosine hydroxylase (TH) expression in the control and MPTP-induced PD mice brains: (a) immunohistochemistry (IHC) images; (b) IHC-fluorescence (IHF) images; (c) Confocal images showing the distributions of ceria NPs in each group. Scale bars: 1 mm. MPTP: 1-methyl-4-phenyl-1,2,3,6-tetrahydropyridine. Reproduced with permission (Kwon et al. 2018). Copyright 2018, Wiley–VCH

3.2 NP-Based Drug Delivery Carriers

The delivery of drugs to the central nervous system (CNS) is highly indispensable and a key challenge in the development of drugs (Dong 2018; Terstappen et al. 2021). Several technologies have been developed to deliver therapeutics or biopharmaceuticals such as monoclonal antibodies to the CNS, some of which have entered clinical trials. To achieve the best and sufficient delivery of drugs across the brain, all physiological barriers (e.g., cellular uptake barrier, BBB, endosomal/lysosomal barrier, and controlled drug release) have to be overcome (Banks 2016; Arvanitis et al. 2020). Different NP based designs for overcoming the above biological barriers to drug delivery has been delineated earlier (Blanco et al. 2015).

Drug delivery vectors such as polymeric NPs, lipid-based NPs, inorganic NPs, extracellular vesicles, and exosomes have been widely developed for the delivery of nucleic acid drugs to the brain (Blanco et al. 2015; Lu et al. 2023). A shuttle peptide, Angiopep-2 that can cross the BBB, has been found to improve the delivery of gold nanorods (Au-NRs) functionalized with PEG to the brain parenchyma (Velasco-Aguirre et al. 2017).

Mesoporous silica NPs were also used as carriers for the co-delivery of plasmid DNA and siRNA for enhancing the generation of dopaminergic neurons from induced pluripotent stem cells (iPSCs) (Chang et al. 2017). Mesoporous silica NPs loaded with ultrasmall cerium oxide have been used as an ROS-responsive and -scavenging nanomedicine for the application of targeted drug delivery system combined with antioxidant therapy (Purikova et al. 2022).

3.2.1 *Apolipoprotein*

Nanostructured lipid carriers have been widely employed as carriers for different neurodegenerative disorders such as Alzheimer's disease, Parkinson's disease, brain cancer, ischemic stroke, and multiple sclerosis (Tapeinos et al. 2017). High-density lipoproteins have been regarded as nature's multifunctional NPs, opening new avenues in drug delivery strategies (Kuai et al. 2016). High-density apolipoprotein (ApoE3-rHDL)-based nanostructures have been designed to bind to Aβ monomers and oligomers with high affinity and degrade them by glial and liver cells (Song et al. 2014). The in vivo AD animal model study showed the attenuated amyloid deposition and microgliosis, and revival of memory loss.

3.2.2 *Dendrimers, Peptides, and Inorganic NPs*

Dendrimers have been explored as powerful building blocks of nanomaterials for crossing BBB and neuronal cells uptake in the CNS disease (Leiro et al. 2018; Santos et al. 2018). Phosphorus dendrimers have been regarded as drugs for neurodegenerative diseases (Caminade 2017). Polymer NPs such as N-isopropylacrylamide: N-tert-butylacrylamide (NiPAM:BAM) showed both drug delivery capability and anti-amyloid properties (Cabaleiro-Lago et al. 2009; Shcharbin et al. 2017).

Finally, nanotechnology enables for traversing the BBB, thereby allowing brain cancer theranostics (Tang et al. 2019). Some of the other notable findings include benzylamide based anti-inflammatory drug conjugates for CNS delivery (Eden et al.

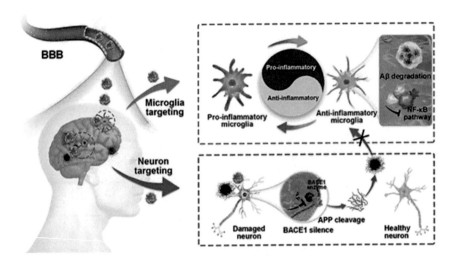

Fig. 3.4 Schematic illustration of lipoprotein-inspired nanoscavenger for modulation of neuroinflammation in Alzheimer's Disease (AD) therapy (Zhang et al. 2022). Copyright 2022 ACS

2019), peptide mediated brain delivery of NPs (McCully et al. 2018), AuNR–peptide–siRNA complex (Vio et al. 2018), fluorescent carbon dots for the targeting of brain cancer cells (Zheng et al. 2015), biodegradable $Cu_{2-x}Se$ NPs for monitoring BBB (Zhang et al. 2018), and stapled RGD peptide for glioma-targeted drug delivery (Ruan et al. 2017).

In a recent interesting work, a lipoprotein-inspired nanoscavenger has been designed to attenuate the inflammatory dysfunction of microglia resulting from excess amyloid-β peptide (Aβ) in Alzheimer's disease (AD) (Zhang et al. 2022). The design consisted of a phosphatidic acid-functionalized high-density lipoprotein (pHDL), curcumin, and β-site APP cleavage enzyme 1 targeted siRNA (siBACE1) to modulate microglial dysfunction, by mimicking the natural lipoprotein transport pathway (Fig. 3.4).

The benefits of this strategy include promoted Aβ clearance (via the pHDL penetration of BBB and sequential targeting of Aβ plaque, an antibody-like Aβ binding affinity), a normalized microglial dysfunction (through blocking of the NF-κB pathway), and reduced Aβ production (via gene silencing, 44%). This method enabled for the reversal of the memory deficit and neuroinflammation in treated mice.

3.3 NP-Based Biosensors for AD Diagnosis

Although therapeutic strategies are important, it is more important to diagnose the AD patients in an early stage using biomarkers (Amen et al. 2022; Oh et al. 2022; Zheng et al. 2023). However, it is a challenge to detect AD blood protein biomarkers owing

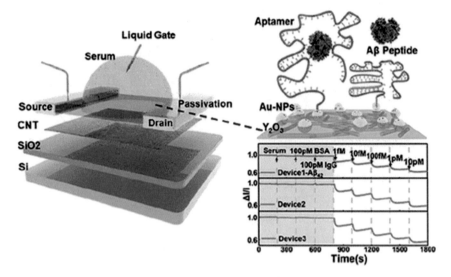

Fig. 3.5 Schematic illustration of a carbon nanotube (CNT) field-effect transistor (FET) biosensor for the selective detection of the AD human serum biomarkers of Aβ42 and Aβ40 peptides in sub-femtomolar detection range (Chen et al. 2022). Copyright 2022, ACS

to a low abundance of biomarkers in a complex serum environment. To obviate these difficulties, recently a nanomaterial-based semiconducting carbon nanotube (CNT) field-effect transistor (FET) biosensor has been designed (Chen et al. 2022). First, thin films of CNT were mass produced and combined with oligonucleotide aptamers to achieve highly sensitive and selective detection of the AD human serum biomarkers of Aβ42 and Aβ40 peptides in sub-femtomolar detection range. The benefits of this approach include the selectivity ratios of up to 730–800% (for Aβ40 and Aβ42), rapid response time (within several minutes), a large dynamic range ($> 10^4$), and low-cost point of care diagnostic test (Fig. 3.5).

Some of the recent advancement in early diagnosis of AD biomarkers include electrochemical aptasensor based on Y probe for the specific molecular recognition of β-amyloid oligomer (Zheng et al. 2023), and several nanotechnology-mediated approaches for both AD and dementia diagnostics and therapy (Leszek et al. 2017; Hettiarachchi et al. 2019; Kumari et al. 2023; Shao et al. 2023).

In conclusion, we have provided a comprehensive overview of the recent developments of nanomedicine assisted strategies (e.g., organic/inorganic NP-based diagnostics and protein-based drug delivery) for inhibiting the amyloid aggregation and tau hyperphosphorylation associated with the Alzheimer's disease (AD), and carbon nanotube-based biosensor strategies for selective detection of the AD human serum biomarkers.

References

Ali T, Kim MJ, Rehman SU, Ahmad A, Kim MO (2017) Anthocyanin-loaded PEG-gold nanoparticles enhanced the neuroprotection of anthocyanins in an Aβ 1–42 mouse model of Alzheimer's disease. Mol Neurobiol 54(8):6490–6506

Ali T, Kim T, Rehman SU, Khan MS, Amin FU, Khan M, Ikram M, Kim MO (2018) Natural dietary supplementation of anthocyanins via PI3K/Akt/Nrf2/HO-1 pathways mitigate oxidative stress, neurodegeneration, and memory impairment in a mouse model of Alzheimer's disease. Mol Neurobiol 55(7):6076–6093

Amen MT, Pham TTT, Cheah E, Tran DP, Thierry B (2022) Metal-oxide FET biosensor for point-of-care testing: overview and perspective. Molecules 27(22):7952

Aparicio-Blanco J, Martin-Sabroso C, Torres-Suarez A-I (2016) In vitro screening of nanomedicines through the blood brain barrier: a critical review. Biomaterials 103:229–255

Arvanitis CD, Ferraro GB, Jain RK (2020) The blood–brain barrier and blood–tumour barrier in brain tumours and metastases. Nat Rev Cancer 20(1):26–41

Banks WA (2016) From blood–brain barrier to blood–brain interface: new opportunities for CNS drug delivery. Nat Rev Drug Discovery 15(4):275–292

Blanco E, Shen H, Ferrari M (2015) Principles of nanoparticle design for overcoming biological barriers to drug delivery. Nat Biotechnol 33(9):941–951

Cabaleiro-Lago C, Lynch I, Dawson K, Linse S (2009) Inhibition of IAPP and IAPP (20–29) fibrillation by polymeric nanoparticles. Langmuir 26(5):3453–3461

Caminade A-M (2017) Phosphorus dendrimers for nanomedicine. Chem Commun 53(71):9830–9838

Chang J-H, Tsai P-H, Chen W, Chiou S-H, Mou C-Y (2017) Dual delivery of siRNA and plasmid DNA using mesoporous silica nanoparticles to differentiate induced pluripotent stem cells into dopaminergic neurons. J Mater Chem B 5(16):3012–3023

Chen Q, Du Y, Zhang K, Liang Z, Li J, Yu H, Ren R, Feng J, Jin Z, Li F (2018) Tau-targeted multifunctional nanocomposite for combinational therapy of Alzheimer's disease. ACS Nano 12(2):1321–1338

Chen H, Xiao M, He J, Zhang Y, Liang Y, Liu H, Zhang Z (2022) Aptamer-functionalized carbon nanotube field-effect transistor biosensors for Alzheimer's disease serum biomarker detection. ACS Sens 7(7):2075–2083

Dong X (2018) Current strategies for brain drug delivery. Theranostics 8(6):1481

Eden BD, Rice AJ, Lovett TD, Toner OM, Geissler EP, Bowman WE, Young SC (2019) Microwave-assisted synthesis and in vitro stability of N-benzylamide non-steroidal anti-inflammatory drug conjugates for CNS delivery. Bioorg Med Chem Lett 29(12):1487–1491

Gao X, Yue Q, Liu Z, Ke M, Zhou X, Li S, Zhang J, Zhang R, Chen L, Mao Y (2017) Guiding brain-tumor surgery via blood–brain-barrier-permeable gold nanoprobes with acid-triggered MRI/SERRS signals. Adv Mater 29(21):1603917

Goldsmith M, Abramovitz L, Peer D (2014) Precision nanomedicine in neurodegenerative diseases. ACS Nano 8(3):1958–1965

Guan Y, Gao N, Ren J, Qu X (2016) Rationally designed CeNP@ MnMoS₄ core-shell nanoparticles for modulating multiple facets of Alzheimer's disease. Chem—Eur J 22(41):14523–14526

Hegazy MA, Maklad HM, Samy DM, Abdelmonsif DA, El Sabaa BM, Elnozahy FY (2017) Cerium oxide nanoparticles could ameliorate behavioral and neurochemical impairments in 6-hydroxydopamine induced Parkinson's disease in rats. Neurochem Int 108:361–371

Hettiarachchi SD, Zhou Y, Seven E, Lakshmana MK, Kaushik AK, Chand HS, Leblanc RM (2019) Nanoparticle-mediated approaches for Alzheimer's disease pathogenesis, diagnosis, and therapeutics. J Control Release 314:125–140

Huang M, Hu M, Song Q, Song H, Huang J, Gu X, Wang X, Chen J, Kang T, Feng X (2015) GM1-modified lipoprotein-like nanoparticle: multifunctional nanoplatform for the combination therapy of Alzheimer's disease. ACS Nano 9(11):10801–10816

Kim CK, Kim T, Choi IY, Soh M, Kim D, Kim YJ, Jang H, Yang HS, Kim JY, Park HK (2012) Ceria nanoparticles that can protect against ischemic stroke. Angew Chem Int Ed 51(44):11039–11043

Kim D, Kwon HJ, Hyeon T (2019) Magnetite/ceria nanoparticle assemblies for extracorporeal cleansing of amyloid-β in Alzheimer's disease. Adv Mater 31(19):1807965

Kuai R, Li D, Chen YE, Moon JJ, Schwendeman A (2016) High-density lipoproteins: nature's multifunctional nanoparticles. ACS Nano 10(3):3015–3041

Kumar A, Tan A, Wong J, Spagnoli JC, Lam J, Blevins BD, Thorne L, Ashkan K, Xie J, Liu H (2017) Nanotechnology for neuroscience: promising approaches for diagnostics, therapeutics and brain activity mapping. Adv Func Mater 27(39):1700489

Kumari P, Das R, Goyal AK (2023) Nanocarriers-based noninvasive approaches for dementia. In: Nanomedicine-based approaches for the treatment of dementia. Elsevier, pp 235-264

Kwon HJ, Cha M-Y, Kim D, Kim DK, Soh M, Shin K, Hyeon T, Mook-Jung I (2016) Mitochondria-targeting ceria nanoparticles as antioxidants for Alzheimer's disease. ACS Nano 10(2):2860–2870

Kwon HJ, Kim D, Seo K, Kim YG, Han SI, Kang T, Soh M, Hyeon T (2018) Ceria nanoparticle systems for selective scavenging of mitochondrial, intracellular, and extracellular reactive oxygen species in Parkinson's disease. Angew Chem Int Ed 57(30):9408–9412

Leiro V, Duque Santos S, Lopes CD, Paula Pêgo A (2018) Dendrimers as powerful building blocks in central nervous system disease: headed for successful nanomedicine. Adv Func Mater 28(12):1700313

Leszek J, Md Ashraf G, Tse WH, Zhang J, Gasiorowski K, Fidel Avila-Rodriguez M, Tarasov VV, Barreto GE, Klochkov SG, Bachurin SO (2017) Nanotechnology for Alzheimer disease. Curr Alzheimer Res 14(11):1182–1189

Lu N-H, How S-C, Lin C-Y, Tsai S-L, Bednarikova Z, Fedunova D, Gazova Z, Wu JW, Wang SS-S (2018) Examining the effects of dextran-based polymer-coated nanoparticles on amyloid fibrillogenesis of human insulin. Colloids Surf, B 172:674–683

Lu Z-G, Shen J, Yang J, Wang J-W, Zhao R-C, Zhang T-L, Guo J, Zhang X (2023) Nucleic acid drug vectors for diagnosis and treatment of brain diseases. Sig Transduct Target Ther 8(1):39

McCully M, Sanchez-Navarro M, Teixido M, Giralt E (2018) Peptide mediated brain delivery of nano-and submicroparticles: a synergistic approach. Curr Pharm Des 24(13):1366–1376

Naz S, Beach J, Heckert B, Tummala T, Pashchenko O, Banerjee T, Santra S (2017) Cerium oxide nanoparticles: a 'radical' approach to neurodegenerative disease treatment. Nanomedicine 12(5):545–553

Oh I-H, Shin W-R, Ahn J, Lee J-P, Min J, Ahn J-Y, Kim Y-H (2022) The present and future of minimally invasive methods for Alzheimer's disease diagnosis. Toxicol Environ Heal Sci 14(4):309–318

Palmal S, Maity AR, Singh BK, Basu S, Jana NR, Jana NR (2014) Inhibition of amyloid fibril growth and dissolution of amyloid fibrils by curcumin–gold nanoparticles. Chem—Eur J 20(20):6184–6191

Purikova O, Tkachenko I, Šmíd B, Veltruská K, Dinhová TN, Vorokhta M, Kopecký V Jr, Hanyková L, Ju X (2022) Free-blockage mesoporous silica nanoparticles loaded with cerium oxide as ROS-responsive and ROS-scavenging nanomedicine. Adv Func Mater 32(46):2208316

Ruan H, Chen X, Xie C, Li B, Ying M, Liu Y, Zhang M, Zhang X, Zhan C, Lu W (2017) Stapled RGD peptide enables glioma-targeted drug delivery by overcoming multiple barriers. ACS Appl Mater Interfaces 9(21):17745–17756

Rzigalinski BA, Carfagna CS, Ehrich M (2017) Cerium oxide nanoparticles in neuroprotection and considerations for efficacy and safety. Wiley Interdisc Rev: Nanomed Nanobiotechnol 9(4):e1444

Santos SD, Xavier M, Leite DM, Moreira DA, Custódio B, Torrado M, Castro R, Leiro V, Rodrigues J, Tomás H (2018) PAMAM dendrimers: blood-brain barrier transport and neuronal uptake after focal brain ischemia. J Control Release 291:65–79

Shao X, Yan C, Wang C, Wang C, Cao Y, Zhou Y, Guan P, Hu X, Zhu W, Ding S (2023) Advanced nanomaterials for modulating Alzheimer's related amyloid aggregation. Nanoscale Adv

Shcharbin D, Shcharbina N, Dzmitruk V, Pedziwiatr-Werbicka E, Ionov M, Mignani S, de la Mata FJ, Gomez R, Muñoz-Fernández MA, Majoral J-P (2017) Dendrimer-protein interactions versus dendrimer-based nanomedicine. Colloids Surf, B 152:414–422

Song Q, Huang M, Yao L, Wang X, Gu X, Chen J, Chen J, Huang J, Hu Q, Kang T (2014) Lipoprotein-based nanoparticles rescue the memory loss of mice with Alzheimer's disease by accelerating the clearance of amyloid-beta. ACS Nano 8(3):2345–2359

Streich C, Akkari L, Decker C, Bormann J, Rehbock C, Müller-Schiffmann A, Niemeyer FC, Nagel-Steger L, Willbold D, Sacca B (2016) Characterizing the effect of multivalent conjugates composed of Aβ-specific ligands and metal nanoparticles on neurotoxic fibrillar aggregation. ACS Nano 10(8):7582–7597

Tamil Selvan S, Padmanabhan P, Zoltán Gulyás B (2020) Nanotechnology-based diagnostics and therapy for pathogen-related infections in the CNS. ACS Chem Neurosci 11(16):2371–2377

Tamil Selvan S, Ravichandar R, Kanta Ghosh K, Mohan A, Mahalakshmi P, Gulyás B, Padmanabhan P (2021) Coordination chemistry of ligands: insights into the design of amyloid beta/tau-PET imaging probes and nanoparticles-based therapies for Alzheimer's disease. Coord Chem Rev 430:213659

Tang W, Fan W, Lau J, Deng L, Shen Z, Chen X (2019) Emerging blood–brain-barrier-crossing nanotechnology for brain cancer theranostics. Chem Soc Rev

Tapeinos C, Battaglini M, Ciofani G (2017) Advances in the design of solid lipid nanoparticles and nanostructured lipid carriers for targeting brain diseases. J Control Release 264:306–332

Terstappen GC, Meyer AH, Bell RD, Zhang W (2021) Strategies for delivering therapeutics across the blood–brain barrier. Nat Rev Drug Discov 20(5):362–383

Velasco-Aguirre C, Morales-Zavala F, Salas-Huenuleo E, Gallardo-Toledo E, Andonie O, Muñoz L, Rojas X, Acosta G, Sánchez-Navarro M, Giralt E (2017) Improving gold nanorod delivery to the central nervous system by conjugation to the shuttle Angiopep-2. Nanomedicine 12(20):2503–2517

Vio V, Riveros AL, Tapia-Bustos A, Lespay-Rebolledo C, Perez-Lobos R, Muñoz L, Pismante P, Morales P, Araya E, Hassan N (2018) Gold nanorods/siRNA complex administration for knockdown of PARP-1: a potential treatment for perinatal asphyxia. Int J Nanomed 13:6839

Wang B, Pilkington EH, Sun Y, Davis TP, Ke PC, Ding F (2017) Modulating protein amyloid aggregation with nanomaterials. Environ Sci Nano 4(9):1772–1783

Xiong N, Zhao Y, Dong X, Zheng J, Sun Y (2017) Design of a molecular hybrid of dual peptide inhibitors coupled on AuNPs for enhanced inhibition of amyloid β-protein aggregation and cytotoxicity. Small 13(13):1601666

Zhang H, Wang T, Qiu W, Han Y, Sun Q, Zeng J, Yan F, Zheng H, Li Z, Gao M (2018) Monitoring the opening and recovery of the blood-brain barrier with noninvasive molecular imaging by biodegradable ultrasmall Cu_{2-x} Se nanoparticles. Nano Lett 18(8):4985–4992

Zhang H, Jiang W, Zhao Y, Song T, Xi Y, Han G, Jin Y, Song M, Bai K, Zhou J, Ding Y (2022) Lipoprotein-inspired nanoscavenger for the three-pronged modulation of microglia-derived neuroinflammation in Alzheimer's disease therapy. Nano Lett 22(6):2450–2460

Zheng M, Ruan S, Liu S, Sun T, Qu D, Zhao H, Xie Z, Gao H, Jing X, Sun Z (2015) Self-targeting fluorescent carbon dots for diagnosis of brain cancer cells. ACS Nano 9(11):11455–11461

Zheng Y, Zhao J, Zhang L, Wang W (2023) Boosting the specific molecular recognition of β-amyloid oligomer by constructing of antibody-mimic Y probe. Sens Actuators B: Chem 133418

Chapter 4
Nanomedicine for Orthopaedics

Abstract This chapter discusses the usefulness of different nanocomposites for the promotion of angiogenesis, osteogenesis, and bone regeneration. Typical examples include graphene oxide–collagen nanocomposite, polylactic acid (PLA)—carbon nanotubes, dexamethasone-loaded calcium phosphate—collagen nanocomposites, nanofiber scaffolds (e.g., chitosan/poly-ethylene oxide), porous hydroxyapatite (HA) and reduced graphene oxide (rGO) nanocomposite scaffold, and injectable bone cement reinforced with Au NPs decorated HA/rGO nanocomposites for bone regeneration.

Keywords Nanomedicine · Bone regeneration · Graphene oxide/collagen nanocomposite · Chitosan/poly-ethylene oxide scaffolds · Porous hydroxyapatite/graphene oxide scaffolds · Hydrogel scaffolds · Tissue engineering · Orthopaedics

Nanomedicine and tissue engineering approaches have manifested great promise in orthopaedic trauma, including high bacterial infection risk, low bony reconstruction and bone-healing (Behzadi et al. 2017). The emergence of novel nanocomposite materials provide great platforms in addressing these issues by exhibiting not only antibacterial properties, but also offer mechanical, biochemical, physicochemical properties, and facilitate osteogenesis/angiogenesis necessary to accelerate the healing process in bone regeneration (Sullivan et al. 2014).

4.1 Nanoscale Materials for Bone Tissue Regeneration

Nanoscale materials such as hydroxyapatite (HA) and calcium phosphate NPs can integrate easily with the bone since they are the main constituent of the natural bone. β-tricalcium phosphate (β-TCP) has both osteo-conductive, osteo-inductive, and cell-mediated resorption properties that promotes the new bone formation due to its excellent biodegradability (degradation into calcium and phosphate ions) in the body (Park et al. 2018).

4.1.1 Natural Polymers Based Scaffolds for Bone Regeneration

Several β-TCP based scaffolds combined with natural polymers, such as chitosan/gelatin (Serra et al. 2015), and dextran (Ghaffari et al. 2020), and interconnected porous scaffolds (Wang et al. 2019a, b) have been demonstrated to be useful for bone tissue regeneration. These scaffolds promoted cell migration, proliferation, and angiogenesis. Similarly, β-TCP based scaffolds combined with graphene oxide (GO) nanoparticles showed great potential in bone tissue engineering (Liu et al. 2020).

Studies have demonstrated the applications of nanocomposites such as polylactic acid (PLA)—carbon nanotubes for enhanced osteoblast adhesion and proliferation (Mazaheri et al. 2015), dexamethasone-loaded calcium phosphate—collagen nanocomposites for the promotion of angiogenesis and osteogenesis (Chen et al. 2018), and nanofiber scaffolds (e.g. chitosan/poly-ethylene oxide) for bone regeneration (Christenson et al. 2007).

Although hydrogels such as type I collagen (COL) has active role in bone tissue repair, it suffers from poor mechanical strength. Addition of graphene oxide (GO) nanosheets greatly enhance the mechanical strength. This has been recently demonstrated in curing the cranial defects in rats by incorporating GO-COL into cultured osteo-differentiated (OiECM) bone marrow mesenchymal stem cells (BMSCs) (Fig. 4.1) (Liu et al. 2018). This approach ameliorated the osteogenic ability, mechanical strength, and biocompatibility of the GO-COL hydrogels. In another recent study, mesoporous silica bioactive glass—graphene oxide scaffold demonstrated its potential in promoting vascular ingrowth, thereby repairing bone defects in a rat cranial model (Wang et al. 2019a, b).

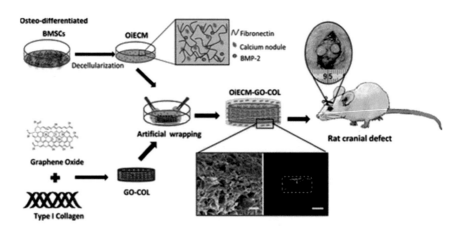

Fig. 4.1 Application of graphene oxide–collagen-I nanocomposite along with osteo-inductive ECM in curing the cranial defect in rat. Reproduced with permission (Liu et al. 2018). Copyright 2018, ACS

4.1.2 Hydroxyapatite and Graphene Oxide Scaffolds for Bone Regeneration

Three-dimensional porous hydroxyapatite and reduced graphene oxide HA/rGO scaffolds with ordered pore structure enabled beneficial cell adhesion and ingrowth, promoting the osteogenic differentiation of BMSCs (Fig. 4.2) (Zhou et al. 2019). The HA/rGO-6/0.3 scaffolds did not show cytotoxicity, but promoted the cell proliferation of BMSCs. This study also confirmed that newly formed bones filled inside the scaffold. With many new bone formations, the scaffold cracked slowly and exposed rGO for further proliferation of stem cells. Finally, the cracked scaffold degraded and wrapped by the new bone. These implants accelerated bone ingrowth and repair of fractured sites. In another interesting work, injectable bone cement was reinforced with Au NPs decorated HA/rGO nanocomposites to strengthen bone regeneration (Chopra et al. 2023).

Other recent work for bone regeneration and anti-infection demonstrated the preparation of nano-HA scaffold loaded with glycopeptide antibiotics (e.g., vancomycin) and its sustained release from the composite polylactic acid (PLA) and poly (lactic acid-glycolic acid) scaffold microspheres (Li et al. 2023). These scaffolds showed excellent biomechanical and biocompatible properties, thereby effectively inhibiting the growth of *Staphylococcus aureus*, and repairing bone defects.

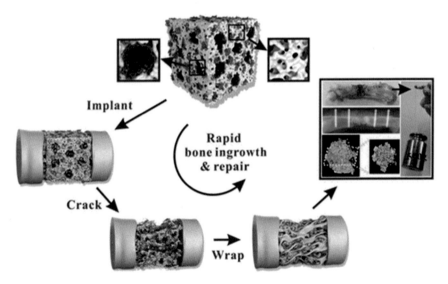

Fig. 4.2 Flow-chart representation of the application of porous scaffold made of hydroxyapatite and reduced graphene oxide nanocomposite for bone ingrowth and repair. Reproduced with permission (Zhou et al. 2019). Copyright 2019, ACS

4.1.3 Antibacterial Nanoparticles Coated Orthopaedic Implants

In orthopaedic implants, NPs have been used especially as coating materials to combat bacterial infections. The drug resistant bacterial infections (e.g., Methicillin-resistant *Staphylococcus aureus*, *Staphylococcus epidermidis*) are serious in traumatic injury patients who stay longer in intensive care units. The selenium (Se) NP (with the size of 30–70 nm) coatings on titanium implants strongly inhibited the biofilm formation, caused by the aforementioned drug resistant bacteria in an infected femur model rats (Tran et al. 2019).

Chitosan scaffolds with Ag or Se NPs served effectively as antibacterial implants for wound dressing application (Biswas et al. 2018). Although both Ag and Se NPs showed their antibacterial effects to three bacterial strains (*S. aureus*, *MRSA and E. coli*), Ag caused cytotoxicity to mouse fibroblast cells, whereas Se NPs did not. Conversely, Au-NPs owing to their anti-inflammatory property, exhibited promising therapeutic effects to iontophoresis, which is a therapeutic treatment of injury. The Au-NPs enhanced the effect of a nonsteroidal anti-inflammatory drug, diclofenac diethylammonium, while reducing inflammatory cytokines in treating traumatic Achilles tendinitis (Dohnert et al. 2012). These three combinations (Au-NP, drug, iontophoresis) lowered the levels of inflammatory cytokines in treated rats, compared to untreated control groups.

The focal skeletal malignant osteolysis is another serious problem. Polymer based NPs such as polylactide formulations loaded with DOX, coated with bone-seeking pamidronate, have been used for the targeted therapy of malignant skeletal tumors (Yin et al. 2016). This nanoformulation attenuated the focal skeletal malignant osteolysis progression in a murine model, compared to the control, non-targeted DOX-NPs.

4.2 Hydrogel Scaffolds for Osteogenic Differentiation of Stem Cells

Novel zwitterionic Chitosan/β-tricalcium phosphate hydrogel/GO scaffolds have also been used for bone tissue engineering (Wang et al. 2023). These scaffolds displayed improved osteogenic differentiation of bone mesenchymal stem cells (Fig. 4.3). The porosity of these scaffolds reported to be increased with an increase in GO concentration. Both swelling and degradation percentage of the scaffolds decreased with increasing GO concentration.

Alternatively, calcium silicate (CS) bioceramics has attracted a great deal of attention, owing to their excellent capability to stimulate osteogenesis (Zhou et al. 2021). This work has demonstrated a significant enhancement of the osteogenic differentiation of bone marrow mesenchymal stem cells (BMSCs) via the stimulation of a macrophage conditioned medium pre-treated with CS extracts. Compared to β-TCP

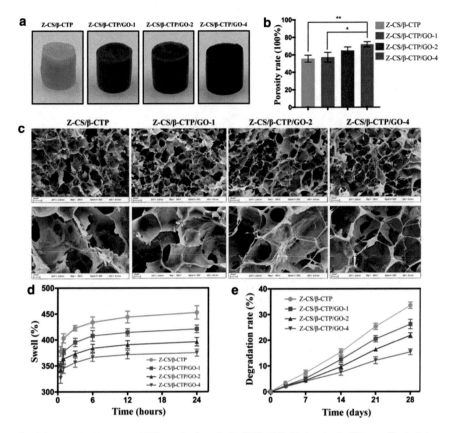

Fig. 4.3 **a** Optical images of different hydrogels. Z-CS/β-CTP: Zwitterionic Chitosan/β-tricalcium phosphate; GO: graphene oxide. The color gets intensified with an increase of GO (GO-1 to GO-4). **b** Porosity of corresponding scaffolds as in (**a**) increased with an increase in GO concentration. **c** SEM images of those scaffolds with different GO concentrations. Lower panels are enlarged images of those red dotted boxes in the upper panels. **d** Swelling percentage and **e** Degradation percentage of scaffolds with different GO concentrations (Wang et al. 2023)

implants, CS scaffolds promoted the osteogenesis due to the presence of oncostatin-M (OSM) protein in the macrophage-conditioned medium, thereby accelerating the new bone formation at defective sites in the femoral bone defects in rats. Micro-CT 3D reconstruction images, and the percentage calculation of bone mineral density (BMD), and bone volume relative to its total tissue volume (BV/TV) confirmed the superiority of CS scaffolds implants over β-TCP implants (Fig. 4.4).

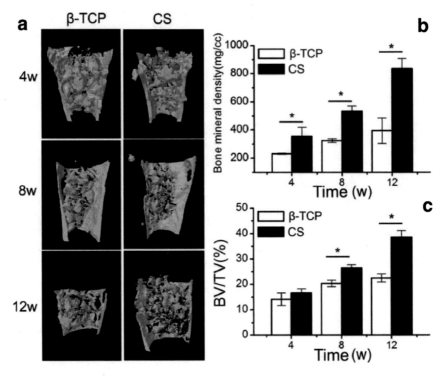

Fig. 4.4 **a** Micro-CT 3D reconstruction (coronal section) of bone regeneration in the femoral bone defect in animals; white, yellow, and blue refer to the original bone, new bone, and implanted material, respectively. **b** Bone mineral density (BMD) versus time in weeks. **c** Percentage of new bone formation, relative to its total tissue volume (BV/TV) (n = 10 rats/batch) (Zhou et al. 2021)

4.2.1 Polymer/Stem Cells Implanted Hydrogels for Cartilage Regeneration

Very recently, an innovative articular cartilage (ARTiCAR) implant combining nanofibrous poly-ε-caprolactone (PCL) and BMSCs in alginate/hyaluronic acid hydrogel has been reported for osteoarticular regeneration (OAR) (Keller et al. 2019). The preclinical safety evaluation of the NanoM1-BMP2 bone wound dressing implant in vitro and BMSCs mixed in alginate/hyaluronic acid hydrogel in vivo have been carried out for subchondral bone and cartilage regeneration (Fig. 4.5).

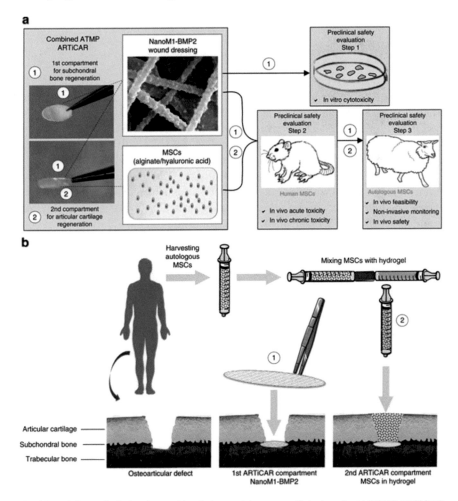

Fig. 4.5 **a** Scheme depicting the combined advanced therapy medicinal product (ATMP) ARTiCAR (ARTicular CArtilage and subchondRal bone implant) involving subchondral bone (compartment 1) and articular cartilage (compartment 2) regenerations. The composite combines an FDA-approved synthetic polymer poly-ε-caprolactone (PCL) and bone marrow-derived mesenchymal stem cells (BMSCs) embedded in alginate/hyaluronic acid hydrogel. The NanoM1-BMP2 refers to the nanofibrous PCL functionalized with bone morphogenetic protein 2 and tested for in vitro cytotoxicity (step 1). The whole ARTiCAR tested for in vivo toxicity, biodistribution studies in an osteochondral defect nude rat model (step 2) and for feasibility, non-invasive monitoring (step 3) in a sheep intra-articular model. **b** Cartoon depicting the usefulness of The ARTiCAR for simultaneous regeneration of the articular cartilage and the subchondral bone. In the first step, the NanoM1-BMP2 is applied to the injured subchondral bone. In the second step, the harvested BMSCs from the patient is mixed with the hydrogel and applied to fill the osteoarticular defect. Reproduced with permission (Keller et al. 2019). Copyright 2019, Nature Publishing Group

4.3 Nanomedicine for Controlled Orthopaedic Drug Delivery

To avoid the complications such as infections associated with post-operative orthopaedic surgery, nanomedicine based controlled drug delivery of antibiotics would be beneficial. Plant based polysaccharides have emerged as potential candidate systems for orthopaedic drug delivery and treatment (Hormozi 2023).

Three-dimensional (3D) printing technology is a powerful technique for the preparation of tissue engineering scaffolds (Adarkwa et al. 2023; Mo et al. 2023; Samie et al. 2023; Song et al. 2023). For instance, polylactic acid (PLA)/nano-hydroxyapatite (nHA) scaffolds (with high porosity and interconnected 3D networks) were loaded with vancomycin (Van)-based chitosan (CS) hydrogel (CS-Van). The resulting PLA/nHA/CS-Van composite scaffold showed enhanced mechanical and biocompatible properties, hydrophilicity, sustained release of the drug in vitro (> 8 weeks). Furthermore, the composite scaffold inhibited the growth of *Staphylococcus aureus* (*S. aureus*) effectively, demonstrating its efficacy for treating infected bone defects (Fig. 4.6) (Gao et al. 2023).

In conclusion, nanomedicine and tissue engineering approaches have shown great promises in orthopaedics. In recent years, 3D bioprinting of bone mimicking scaffolds comprised of polymer (e.g., polylactic acid, polycaprolactone), and inorganic NPs (e.g., reduced graphene oxide, hydroxyapatite/chitosan) have emerged as potential candidate systems for bone tissue engineering (Seyedsalehi et al. 2020; Zhang et al. 2021). This chapter covered the applicability of these scaffolds in promoting scaffold fidelity, osteogenic differentiation and mineralization.

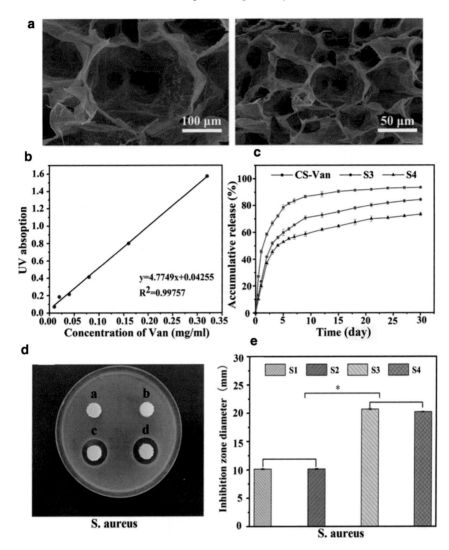

Fig. 4.6 **a** SEM images of CS-Van, **b** the standard curve of Van, **c** In vitro release profiles of CS-Van, S3 and S4, **d** antibacterial test, **e** inhibition zone diameter for *S. aureus* with (a) PLA/nHA vertical orthogonal scaffold (S1), (b) PLA/nHA staggered orthogonal scaffold (S2), (c) PLA/nHA/CS-Van vertical orthogonal scaffold (S3) and (d) PLA/nHA/CS-Van staggered orthogonal scaffold (S4) (Gao et al. 2023) Copyright 2023, RSC

References

Adarkwa E, Roy A, Ohodnicki J, Lee B, Kumta PN, Desai S (2023) 3D printing of drug-eluting bioactive multifunctional coatings for orthopedic applications. Int J Bioprint 9(2)

Behzadi S, Luther GA, Harris MB, Farokhzad OC, Mahmoudi M (2017) Nanomedicine for safe healing of bone trauma: opportunities and challenges. Biomaterials 146:168–182

Biswas DP, O'Brien-Simpson NM, Reynolds EC, O'Connor AJ, Tran PA (2018) Comparative study of novel in situ decorated porous chitosan-selenium scaffolds and porous chitosan-silver scaffolds towards antimicrobial wound dressing application. J Colloid Interface Sci 515:78–91

Chen Y, Chen S, Kawazoe N, Chen G (2018) Promoted angiogenesis and osteogenesis by dexamethasone-loaded calcium phosphate nanoparticles/collagen composite scaffolds with microgroove networks. Sci Rep 8(1):14143

Chopra V, Thomas J, Kaushik S, Rajput S, Guha R, Mondal B, Naskar S, Mandal D, Chauhan G, Chattopadhyay N (2023) Injectable bone cement reinforced with gold nanodots decorated rGO-hydroxyapatite nanocomposites, augment bone regeneration. Small 2204637

Christenson EM, Anseth KS, van den Beucken JJ, Chan CK, Ercan B, Jansen JA, Laurencin CT, Li WJ, Murugan R, Nair LS (2007) Nanobiomaterial applications in orthopedics. J Orthop Res 25(1):11–22

Dohnert MB, Venâncio M, Possato JC, Zeferino RC, Dohnert LH, Zugno AI, De Souza CT, Paula MM, Luciano TF (2012) Gold nanoparticles and diclofenac diethylammonium administered by iontophoresis reduce inflammatory cytokines expression in *Achilles tendinitis*. Int J Nanomed 7:1651

Gao X, Xu Z, Li S, Cheng L, Xu D, Li L, Chen L, Xu Y, Liu Z, Liu Y (2023) Chitosan-vancomycin hydrogel incorporated bone repair scaffold based on staggered orthogonal structure: a viable dually controlled drug delivery system. RSC Adv 13(6):3759–3765

Ghaffari R, Salimi-Kenari H, Fahimipour F, Rabiee SM, Adeli H, Dashtimoghadam E (2020) Fabrication and characterization of dextran/nanocrystalline β-tricalcium phosphate nanocomposite hydrogel scaffolds. Int J Biol Macromol 148:434–448

Hormozi B (2023) Plant polysaccharides for orthopedic drug delivery. Plant polysaccharides as pharmaceutical excipients. Elsevier, pp 513–532

Keller L, Pijnenburg L, Idoux-Gillet Y, Bornert F, Benameur L, Tabrizian M, Auvray P, Rosset P, Gonzalo-Daganzo RM, Barrena EG (2019) Preclinical safety study of a combined therapeutic bone wound dressing for osteoarticular regeneration. Nat Commun 10

Li J, Li K, Du Y, Tang X, Liu C, Cao S, Zhao B, Huang H, Zhao H, Kong W (2023) Dual-nozzle 3D printed nano-hydroxyapatite scaffold loaded with vancomycin sustained-release microspheres for enhancing bone regeneration. Int J Nanomed 307–322

Liu S, Mou S, Zhou C, Guo L, Zhong A, Yang J, Yuan Q, Wang J, Sun J, Wang Z (2018) Off-the-shelf biomimetic graphene oxide-collagen hybrid scaffolds wrapped with osteoinductive extracellular matrix for the repair of cranial defects in rats. ACS Appl Mater Interfaces 10(49):42948–42958

Liu F, Liu C, Zheng B, He J, Liu J, Chen C, Lee I-S, Wang X, Liu Y (2020) Synergistic effects on incorporation of β-tricalcium phosphate and graphene oxide nanoparticles to silk fibroin/soy protein isolate scaffolds for bone tissue engineering. Polymers 12(1):69

Mazaheri M, Eslahi N, Ordikhani F, Tamjid E, Simchi A (2015) Nanomedicine applications in orthopedic medicine: state of the art. Int J Nanomed 10:6039

Mo X, Zhang D, Liu K, Zhao X, Li X, Wang W (2023) Nano-hydroxyapatite composite scaffolds loaded with bioactive factors and drugs for bone tissue engineering. Int J Mol Sci 24(2):1291

Park H, Kim JS, Oh EJ, Kim TJ, Kim HM, Shim JH, Yoon WS, Huh JB, Moon SH, Kang SS (2018) Effects of three-dimensionally printed polycaprolactone/β-tricalcium phosphate scaffold on osteogenic differentiation of adipose tissue-and bone marrow-derived stem cells. Arch Craniofac Surg 19(3):181

Samie M, Khan AF, Rahman SU, Iqbal H, Yameen MA, Chaudhry AA, Galeb HA, Halcovitch NR, Hardy JG (2023) Drug/bioactive eluting chitosan composite foams for osteochondral tissue engineering. Int J Biol Macromol 229:561–574

Serra I, Fradique R, M. C. d. S. Vallejo, T. R. Correia, S. P. Miguel and I. Correia, (2015) Production and characterization of chitosan/gelatin/β-TCP scaffolds for improved bone tissue regeneration. Mater Sci Eng C 55:592–604

Seyedsalehi A, Daneshmandi L, Barajaa M, Riordan J, Laurencin CT (2020) Fabrication and characterization of mechanically competent 3D printed polycaprolactone-reduced graphene oxide scaffolds. Sci Rep 10(1):22210

Song Y, Hu Q, Liu Q, Liu S, Wang Y, Zhang H (2023) Design and fabrication of drug-loaded alginate/hydroxyapatite/collagen composite scaffolds for repairing infected bone defects. J Mater Sci 1–16

Sullivan M, McHale K, Parvizi J, Mehta S (2014) Nanotechnology: current concepts in orthopaedic surgery and future directions. Bone Joint J 96(5):569–573

Tran PA, O'Brien-Simpson N, Palmer JA, Bock N, Reynolds EC, Webster TJ, Deva A, Morrison WA, O'Connor AJ (2019) Selenium nanoparticles as anti-infective implant coatings for trauma orthopedics against methicillin-resistant *Staphylococcus aureus* and epidermidis: in vitro and in vivo assessment. Int J Nanomed 14:4613

Wang W, Liu Y, Yang C, Qi X, Li S, Liu C, Li X (2019a) Mesoporous bioactive glass combined with graphene oxide scaffolds for bone repair. Int J Biol Sci 15(10):2156–2169

Wang X, Lin M, Kang Y (2019b) Engineering porous β-tricalcium phosphate (β-TCP) scaffolds with multiple channels to promote cell migration, proliferation, and angiogenesis. ACS Appl Mater Interfaces 11(9):9223–9232

Wang Q, Li M, Cui T, Wu R, Guo F, Fu M, Zhu Y, Yang C, Chen B, Sun G (2023) A novel zwitterionic hydrogel incorporated with graphene oxide for bone tissue engineering: synthesis, characterization, and promotion of osteogenic differentiation of bone mesenchymal stem cells. Int J Mol Sci 24(3):2691

Yin Q, Tang L, Cai K, Tong R, Sternberg R, Yang X, Dobrucki LW, Borst LB, Kamstock D, Song Z (2016) Pamidronate functionalized nanoconjugates for targeted therapy of focal skeletal malignant osteolysis. Proc Natl Acad Sci 113(32):E4601–E4609

Zhang J, Eyisoylu H, Qin X-H, Rubert M, Müller R (2021) 3D bioprinting of graphene oxide-incorporated cell-laden bone mimicking scaffolds for promoting scaffold fidelity, osteogenic differentiation and mineralization. Acta Biomater 121:637–652

Zhou K, Yu P, Shi X, Ling T, Zeng W, Chen A, Yang W, Zhou Z (2019) Hierarchically porous hydroxyapatite hybrid scaffold incorporated with reduced graphene oxide for rapid bone ingrowth and repair. ACS Nano 13(8):9595–9606

Zhou P, Xia D, Ni Z, Ou T, Wang Y, Zhang H, Mao L, Lin K, Xu S, Liu J (2021) Calcium silicate bioactive ceramics induce osteogenesis through oncostatin M. Bioact Mater 6(3):810–822

Chapter 5
Nanomedicine for Cardiac Diseases

Abstract Nanomedicine is a rapidly developing field with the potential to transform the treatment of cardiac diseases. It entails the use of nanoscale materials and devices to diagnose, treat, and prevent a variety of cardiovascular diseases. The development of nanocarriers for drug delivery is a key application of nanomedicine in cardiology. This chapter delineates the applications of nanomedicine in diagnostics and treatment of cardiovascular diseases.

Keywords Nanomedicine · Cardiovascular diseases · Dual drug delivery · Dextran NPs · Cardiac tissue repair · Nanoparticle/composite patch · Exosomes · Stem cells · Cardiac regeneration

Cardiovascular diseases or disorders (CVDs) are one of the leading causes of significant morbidity and mortality. The increasing mortality rate is often attributed to hypertension, a key factor for the onset of CVDs such as myocardial infarction, atherosclerosis, thrombosis, and restenosis. Among these CVDs, atherosclerosis (thickening of the arterial vessel wall due to build-up of fats, cholesterol), myocardial infarction (heart attack), coronary artery disease, and arrhythmias (irregular heartbeat) are considered as the main causes of heart failure. This requires the development of novel diagnostic and therapeutic methods for chronic heart failure (Haeck et al. 2012). Although considerable progress has been made over the years in imaging modalities such as magnetic resonance imaging (MRI) and computed tomography (CT), diagnostic nanomedicine aims to improve the early detection by using nanoparticle-based contrast agents for better visualization of the heart and blood vessels.

5.1 Nanomedicine for Diagnostics and Therapeutics of CVDs

Nanomedicine has a great potential for the diagnostics and treatment of CVDs. Nanomedicine based therapeutics and molecular imaging methods emerge as novel strategies in addressing the morbidity and mortality associated with myocardial infarction (Ferreira et al. 2015). Nanomedicine enables controlled delivery of active drugs encapsulated in nanocarriers for targeted delivery into the vascular site. This targeted nanomedicine approach is increasingly used for the treatment of CVDs like the dissolution of atherosclerotic plaques accumulated in the coronary arteries walls (Kleinstreuer et al. 2018).

5.1.1 Nanomedicine for Cardiac Tissue Repair and Reinforcement

Importantly, nano/biomaterials combined with stem cells offer great promises for cardiac tissue repair and reinforcement (Hasan et al. 2016; Tariq et al. 2022). Direct delivery of cardioprotective drugs (e.g., adenosine) into the ischemic-reperfused myocardium poses great challenges. Biodegradable silica nanoparticles have been used as carriers for the delivery of adenosine into ischemic-reperfused heart tissue (Galagudza et al. 2012).

Recently, several reviews focused on addressing novel nanoparticles (NPs) and nanomedicine approaches for early disease diagnosis (e.g., molecular imaging) and advanced therapeutic (e.g., drug eluting stents) applications of CVDs such as restenosis, atherosclerosis, and MI (Godin et al. 2010; Chopra et al. 2022; Mohamed et al. 2022; Ouyang et al. 2022; Saeed et al. 2023).

Oxidative stress induced free radicals are closely associated with atherosclerosis and many other heart diseases. Polymeric PLGA NPs loaded with an antioxidant drug quercetin, improved the bioavailability of the drug, and prevented the atherosclerosis (Giannouli et al. 2018). Biodegradable porous silicon NPs functionalized with atrial natriuretic peptide and loaded with a cardioprotective small molecule reduced the hypertrophic signaling in the endocardium, demonstrating the targeted delivery of drug to the injured region of the myocardium (Ferreira et al. 2017).

5.1.2 Nanocarriers for Cardiac Delivery of Drugs and MicroRNA

Calcium phosphate NPs have been emerged as therapeutic vehicles for cardiac delivery of MicroRNAs (Di Mauro et al. 2016). Studies have shown that calcium phosphate NPs can be internalized and delivered microRNAs efficiently into cardiac cells (cardiomyocytes) both in vitro and in vivo.

Interestingly, NPs have also been used as protective agents to minimize the cardiac toxicity and disorders associated with oxidative stress. For instance, cerium oxide NPs showed a promising ameliorative and prophylactic toxicity effect, compared to the reference drug, Captopril (El Shaer et al. 2017). Vesicles such as exosomes (< 100 nm) could serve as vehicles for lipids, proteins, DNA and RNAs, and therefore, offers new insights into personalized nanomedicine for acute cardiac, lung, and kidney injury (Terrasini and Lionetti 2017).

Nanotechnology-based diagnostics and therapeutics using mesoporous silica NPs have been used to study the heart failure in cardiac tissue of a murine heart failure model. After intravenous administration, these nanovectors were found to be internalized and accumulated in failing myocardium, through the labelling of perinuclear region of cardiomyocytes in vivo (Ruiz-Esparza et al. 2016).

Other diseases and their treatment methods are closely associated to heart failure. For instance, anticancer drugs are prone to exhibit toxicity to cardiac cells. However, when DOX is encapsulated within exosomes, no cardiac toxicity observed from in vivo studies, resulting in about 40% of accumulation of exosome-DOX in the heart (Toffoli et al. 2015). The detoxifying action of liposomes was demonstrated in a mouse cardiovascular model (Bertrand et al. 2010).

Nanomedicine offers great potential in cardiac catheterization via targeted drug delivery, while preventing the endothelial dysfunction or damage of cells by reduced inflammation or increased nitric oxide bioavailability (Sobolewski and El Fray 2015). The sustained delivery of cardiac therapeutics [Pyr1]-apelin-13 polypeptide in vivo for treating heart injuries was achieved by PEG-conjugated liposomal NPs. The cardiac dysfunction was prevented by this approach through the sustained bioavailability of cardio-protective therapeutics (Serpooshan et al. 2015).

5.2 Nanotechnology Assisted Cardiac Regenerative Medicine

Cardiac regenerative medicine is an emerging paradigm. Carbon nanotubes (CNTs) have received considerable interest in cardiovascular system, owing to its remarkable characteristics in promoting in vitro growth of cardiac cells and improving proliferation, maturation, and electrical behaviour of cardiomyocytes (Martinelli et al. 2013). Myocardial tissues cultured on the CNT–GelMA (gelatin methacrylate) hydrogel scaffold showed high mechanical strength and electro-activity with spontaneous synchronous beating rates, compared to those cardiac tissues cultured on pristine GelMA hydrogels (Shin et al. 2013).

Magnetic bifunctional NPs conjugated with antibodies were used for treating acute myocardial infarction through the targeting of CD45 expressing stem cells (exogenous bone marrow-derived) or CD34-positive cells (endogenous injured cardiomyocytes) (Cheng et al. 2014). Methacrylated gelatin biocompatible hydrogel was used as a delivery vehicle to deliver polyethylenimine (PEI) functionalized graphene oxide nanosheets complexed with a pro-angiogenic gene—vascular endothelial growth factor (VEGF) for myocardial therapy (Paul et al. 2014). This nanocomposite approach showed significant cardiac performance in echocardiography after 14 days of post-injection. The nanocomposite treated infarcted hearts showed a reduction in scar area.

Early allograft acute rejection in heart transplantation is a big problem. Simultaneous diagnosis by non-invasive MRI and gene therapy (mediating gene transfection in T cells) by multifunctional polymeric nanocarriers provided a great relief in heart-transplanted rats (Guo et al. 2012). Nanopatterned (nanofibrous electro-spun) cardiac patches prevented defective electro-coupling and improved the therapeutic efficacy of myocardial infarction, thereby opening avenues in translational myocardial tissue engineering and nanomedicine (Lin et al. 2014).

Cobalt protoporphyrin encapsulated amine-functionalized mesoporous silica NPs (CoPP@aMSNs)—labelled bone marrow stromal cells (BMSCs) have been synthesized (Fig. 5.1) and exhibited photoacoustic imaging (PA) enhancement—guided cell delivery and antioxidant protection of stem cells. (Yao et al. 2018) The drug, CoPP released upon MSN degradation, and endowed the labelled BMSCs with persistent antioxidant activity.

The mouse myocardium implanted with NPs -labelled BMSCs using an insulin needle and analysed by both ultrasound and photoacoustic imaging (Fig. 5.2a–d). The photoacoustic imaging observed at the laser excitation of 680 nm revealed clearly the implanted stem cells (Fig. 5.2a, b). The histological hematoxylin and eosin (H&E) staining of the dissected myocardium after 24 h post-injection showed the engrafted BMSCs corresponding to the enhanced photoacoustic signal region (Fig. 5.2c). Nevertheless, the photoacoustic imaging (Fig. 5.2e, f) showed the leakage of implanted BMSCs into pericardial cavity, indicating the unsuccessful intra-myocardial implantation.

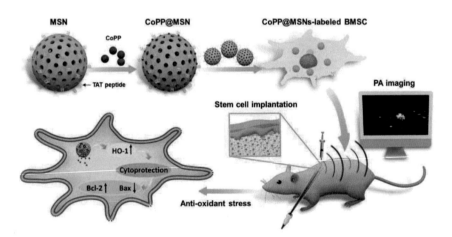

Fig. 5.1 Schematic representation of cobalt protoporphyrin encapsulated mesoporous silica NPs (CoPP@aMSNs)-labelled BMSCs for implantation and cell protected therapy guided by photoacoustic (PA) imaging. TAT peptide facilitates the cell internalization. Reproduced with permission (Yao et al. 2018). Copyright 2018, Wiley-VCH

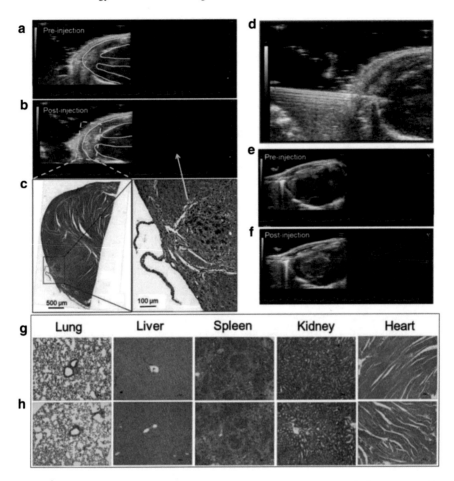

Fig. 5.2 The ultrasound and photoacoustic images of the cardiac structure before (**a**) and after (**b**) intra-myocardial injection of BMSCs. The white curves in the ultrasound images denote the outline of the myocardium, whereas the yellow arrow indicates the strengthened photoacoustic signal from the injected particles. The line patterns at the bottom show the ECG and respiratory coupling signals of nude mouse. **c** Haemotoxylin and Eosin (H&E) stained microscopic image of the BMSCs (blue) in the myocardium. **d** The ultrasound image process of the intramyocardial injection using 30 G needle. **e, f** The long-axis view of cardiac structures before (**e**) and after (**f**) intramyocardial injection of the labelled BMSCs with the signal (dotted circle, **f**) corresponding to the leaked BMSCs in the pericardial cavity. The H&E staining of different organ sections of nude mice for the control (**g**: Matrigel/saline solution) and labelled BMSCs (**h**) taken on day 28. Reproduced with permission (Yao et al. 2018). Copyright 2018, Wiley-VCH

On a positive note, CoPP@aMSNs-labelled BMSCs did not cause any acute toxicity and pathological abnormalities to all major organs (liver, spleen, heart, lung and kidney) after 4 weeks of intra-myocardial injection, indicting the biosafety of this nanomedicine strategy (Fig. 5.2g, h).

5.2.1 Dextran NPs Assisted Dual Drug Delivery for Cardiac Regeneration

Cardiac regeneration holds great promise in restoring the full functionality of a damaged heart. However, it remains a challenge as the injured or damaged heart contains a vast number of fibroblasts and myofibroblasts. To replenish the lost cardiomyocytes, a direct fibroblast cell reprogramming into cardiomyocytes is regarded as an attractive therapeutic option. For this purpose, two small drug molecules, CHIR99021 (aminopyrimidine derivative for glycogen synthase kinase, enzyme GSK-3 inhibitor) and SB-431542 (a drug developed by GlaxoSmithKline for activin receptor-like kinase, ALK5, ALK4 and ALK7 inhibitor) were encapsulated into dextran NPs (functionalized with polyethylene glycol and atrial natriuretic peptide for pH-triggered drug delivery into the affected areas for direct reprogramming of fibroblast into cardiomyocytes (Ferreira et al. 2018).

5.2.2 Nanotherapeutic Approach for Alleviating Cardiac Brain Injury

Another important issue is cardiac arrest/cardiopulmonary resuscitation (CA/CPR)-induced brain injury, which necessitates feasible therapeutic options. To address this, a nanotherapeutic approach based on octanoic acid (OA) and a neutrophil membrane expressing RVG29, RVG29-H-NP-OA has recently been developed for ameliorating the CA/CPR-induced brain injury (Yang et al. 2023). Peptide (RVG29-rabies virus glycoprotein) conjugated OA-NP traversed BBB and targeted the injured brain. Due to their antioxidant, mitochondria stability and anti-inflammatory effects of OA, RVG29-H-NP-OA significantly improved the survival rate and neurological functions of CA/CPR model rats from 40% to 100% over 24 h (Fig. 5.3).

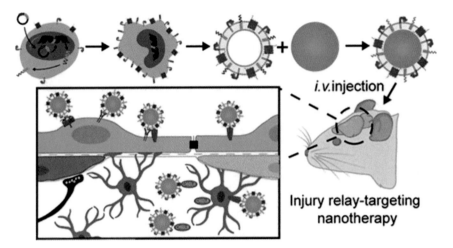

Fig. 5.3 A schematic representation of brain injury relay targeting nanotherapeutic approach based on octanoic acid (OA) and a neutrophil membrane expressing RVG29 peptides for improving the cardiac arrest/cardiopulmonary resuscitation-induced brain injury (Yang et al. 2023). Copyright 2023, ACS

5.2.3 Nanoparticle-Based Composite Patch for Myocardial Infarction

Myocardial infarction, generally identified as heart attack, occurs when blood flow to one or more areas of the heart muscle is blocked. Coronary artery disease is the main cause of myocardial infarction. Cardiac delivery of therapeutic agent may serve as a promising platform for the treatment of heart diseases. Toward this end, nanocarriers have been developed for efficient delivery of therapeutic agents to target the heart. Earlier, porous silicon-based nanomaterials have been used for both diagnostic and therapeutic applications (Li et al. 2018; Tieu et al. 2019). Although biocompatible porous silicon based multifunctional drug delivery systems (DDS) have been widely used for cancer therapy (Zhang et al. 2019), their application in cardiac tissue engineering is limited. Porous silicon micro and nanoparticles showed in vivo biocompatibility to the heart tissue (Tölli et al. 2014).

Recently, a biodegradable polymer (polyglycerol sebacate) nanoparticle conductive composite patch loaded with a small molecule (3i-1000) drug, has been developed for treating myocardial infarction (Zanjanizadeh Ezazi et al. 2020). This nanocomposite contained collagen type I to promote cell attachment, and polypyrrole for inducing electrical conductivity and cell signaling (Fig. 5.4).

Cardiomyoblast cell attachment and morphology on the surface of cardiac patches were observed using a scanning electron microscope (SEM) (Fig. 5.5). High infiltration and attachment of cardiomyoblast cells were found on the collagen-containing patches (0.5C–5P) within 24 h.

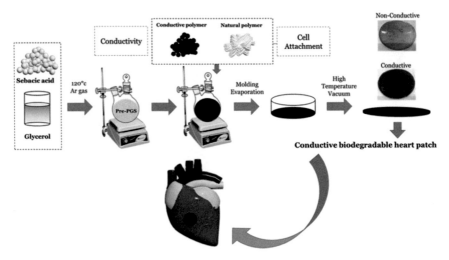

Fig. 5.4 A schematic representation of the synthesis of elastic biodegradable and conductive cardiac patches. Polycondensation of sebacic acid and glycerol was used to prepare polyglycerol sebacate (PGS). Natural polymer (collagen type I) and conductive polymer (polypyrrole PPy) were added. The composite polymer was cured at high temperature under vacuum, resulting in a conductive biodegradable heart patch (Zanjanizadeh Ezazi et al. 2020). Copyright 2020, ACS

The in vitro experiments confirmed that a high density of cardiac myoblast cells was attached on the patches, which remained viable for > 1 month. Conductive patches showed high drug release with no cytotoxic effect upon degradation of the patches, and cell proliferation was induced by the small molecule drug.

A recent work on magnetic-guided accumulation of exosomes suggested that the antibody-conjugated magnetic NPs can be used to capture and deliver circulating CD63-expressing exosomes in infarcted heart tissue, leading to reductions in infarct size as well as improved angiogenesis in rat models of myocardial infarction (Liu et al. 2020).

In conclusion, this chapter reviewed the recent developments of nano/biomaterials having the attributes of mechanical, conductive, and biological requirements for a successful cardiac treatment.

Non-Conductive   Conductive

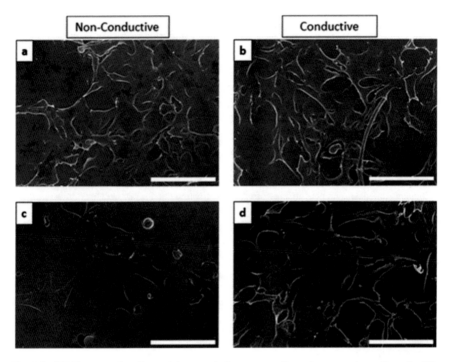

Fig. 5.5 SEM images of cardiomyoblast morphology upon cell attachment on the surface of **a** 0C–0P, **b** 0C–1P, **c** 0.5C–0P, and **d** 0.5C–5P patches on the control (non-conductive) and conductive samples during 24 h. Collagen-containing patches showed high infiltration and attachment on 0.5C–5P patches within 24 h. Scale bar: 100 μm (Zanjanizadeh Ezazi et al. 2020). Copyright 2020, ACS

References

Bertrand N, Bouvet C, Moreau P, Leroux J-C (2010) Transmembrane pH-gradient liposomes to treat cardiovascular drug intoxication. ACS Nano 4(12):7552–7558

Cheng K, Shen D, Hensley MT, Middleton R, Sun B, Liu W, De Couto G, Marbán E (2014) Magnetic antibody-linked nanomatchmakers for therapeutic cell targeting. Nat Commun 5:4880

Chopra H, Bibi S, Mishra AK, Tirth V, Yerramsetty SV, Murali SV, Ahmad SU, Mohanta YK, Attia MS, Algahtani A (2022) Nanomaterials: a promising therapeutic approach for cardiovascular diseases. J Nanomater 2022:1–25

Di Mauro V, Iafisco M, Salvarani N, Vacchiano M, Carullo P, Ramírez-Rodríguez GB, Patrício T, Tampieri A, Miragoli M, Catalucci D (2016) Bioinspired negatively charged calcium phosphate nanocarriers for cardiac delivery of MicroRNAs. Nanomedicine 11(8):891–906

El Shaer SS, Salaheldin TA, Saied NM, Abdelazim SM (2017) In vivo ameliorative effect of cerium oxide nanoparticles in isoproterenol-induced cardiac toxicity. Exp Toxicol Pathol 69(7):435–441

Ferreira MP, Ranjan S, Kinnunen S, Correia A, Talman V, Mäkilä E, Barrios-Lopez B, Kemell M, Balasubramanian V, Salonen J (2017) Drug-loaded multifunctional nanoparticles targeted to the endocardial layer of the injured heart modulate hypertrophic signaling. Small 13(33):1701276

Ferreira PA, M., V. Balasubramanian, J. Hirvonen, H. Ruskoaho and H. A Santos, (2015) Advanced nanomedicines for the treatment and diagnosis of myocardial infarction and heart failure. Curr Drug Targets 16(14):1682–1697

Ferreira MP, Talman V, Torrieri G, Liu D, Marques G, Moslova K, Liu Z, Pinto JF, Hirvonen J, Ruskoaho H (2018) Dual-drug delivery using dextran-functionalized nanoparticles targeting cardiac fibroblasts for cellular reprogramming. Adv Funct Mater 28(15):1705134

Galagudza M, Korolev D, Postnov V, Naumisheva E, Grigorova Y, Uskov I, Shlyakhto E (2012) Passive targeting of ischemic-reperfused myocardium with adenosine-loaded silica nanoparticles. Int J Nanomed 1671–1678

Giannouli M, Karagkiozaki V, Pappa F, Moutsios I, Gravalidis C, Logothetidis S (2018) Fabrication of quercetin-loaded PLGA nanoparticles via electrohydrodynamic atomization for cardiovascular disease. Mater Today Proc 5(8):15998–16005

Godin B, Sakamoto JH, Serda RE, Grattoni A, Bouamrani A, Ferrari M (2010) Emerging applications of nanomedicine for the diagnosis and treatment of cardiovascular diseases. Trends Pharmacol Sci 31(5):199–205

Guo Y, Chen W, Wang W, Shen J, Guo R, Gong F, Lin S, Cheng D, Chen G, Shuai X (2012) Simultaneous diagnosis and gene therapy of immuno-rejection in rat allogeneic heart transplantation model using a T-cell-targeted theranostic nanosystem. ACS Nano 6(12):10646–10657

Haeck ML, Hoogslag GE, Rodrigo SF, Atsma DE, Klautz R, van der Wall EE, Schalij MJ, Verwey HF (2012) Treatment options in end-stage heart failure: where to go from here? Neth Heart J 20:167–175

Hasan A, Waters R, Roula B, Dana R, Yara S, Alexandre T, Paul A (2016) Engineered biomaterials to enhance stem cell-based cardiac tissue engineering and therapy. Macromol Biosci 16(7):958–977

Kleinstreuer C, Chari SV, Vachhani S (2018) Potential use of multifunctional nanoparticles for the treatment of cardiovascular diseases. J Cardiol Cardiovasc Sci 2(3)

Li W, Liu Z, Fontana F, Ding Y, Liu D, Hirvonen JT, Santos HA (2018) Tailoring porous silicon for biomedical applications: from drug delivery to cancer immunotherapy. Adv Mater 30(24):1703740

Lin Y-D, Ko M-C, Wu S-T, Li S-F, Hu J-F, Lai Y-J, Hans I, Harn C, Laio I-C, Yeh M-L (2014) A nanopatterned cell-seeded cardiac patch prevents electro-uncoupling and improves the therapeutic efficacy of cardiac repair. Biomater Sci 2(4):567–580

Liu S, Chen X, Bao L, Liu T, Yuan P, Yang X, Qiu X, Gooding JJ, Bai Y, Xiao J (2020) Treatment of infarcted heart tissue via the capture and local delivery of circulating exosomes through antibody-conjugated magnetic nanoparticles. Nat Biomed Eng 4(11):1063–1075

Martinelli V, Cellot G, Fabbro A, Bosi S, Mestroni L, Ballerini L (2013) Improving cardiac myocytes performance by carbon nanotubes platforms. Front Physiol 4:239

Mohamed NA, Marei I, Crovella S, Abou-Saleh H (2022) Recent developments in nanomaterials-based drug delivery and upgrading treatment of cardiovascular diseases. Int J Mol Sci 23(3):1404

Ouyang J, Xie A, Zhou J, Liu R, Wang L, Liu H, Kong N, Tao W (2022) Minimally invasive nanomedicine: nanotechnology in photo-/ultrasound-/radiation-/magnetism-mediated therapy and imaging. Chem Soc Rev 51(12):4996–5041

Paul A, Hasan A, Kindi HA, Gaharwar AK, Rao VT, Nikkhah M, Shin SR, Krafft D, Dokmeci MR, Shum-Tim D (2014) Injectable graphene oxide/hydrogel-based angiogenic gene delivery system for vasculogenesis and cardiac repair. ACS Nano 8(8):8050–8062

Ruiz-Esparza GU, Segura-Ibarra V, Cordero-Reyes AM, Youker KA, Serda RE, Cruz-Solbes AS, Amione-Guerra J, Yokoi K, Kirui DK, Cara FE (2016) A specifically designed nanoconstruct associates, internalizes, traffics in cardiovascular cells, and accumulates in failing myocardium: a new strategy for heart failure diagnostics and therapeutics. Eur J Heart Fail 18(2):169–178

Saeed S, Khan SU, Gul R (2023) Nanoparticle: a promising player in nanomedicine and its theranostic applications for the treatment of cardiovascular diseases. Curr Probl Cardiol 101599

Serpooshan V, Sivanesan S, Huang X, Mahmoudi M, Malkovskiy AV, Zhao M, Inayathullah M, Wagh D, Zhang XJ, Metzler S (2015) [Pyr1]-Apelin-13 delivery via nano-liposomal encapsulation attenuates pressure overload-induced cardiac dysfunction. Biomaterials 37:289–298

Shin SR, Jung SM, Zalabany M, Kim K, Zorlutuna P, Kim SB, Nikkhah M, Khabiry M, Azize M, Kong J (2013) Carbon-nanotube-embedded hydrogel sheets for engineering cardiac constructs and bioactuators. ACS Nano 7(3):2369–2380

Sobolewski P, El Fray M (2015) Cardiac catheterization: consequences for the endothelium and potential for nanomedicine. Wiley Interdisc Rev Nanomed Nanobiotechnol 7(3):458–473

Tariq U, Gupta M, Pathak S, Patil R, Dohare A, Misra SK (2022) Role of biomaterials in cardiac repair and regeneration: therapeutic intervention for myocardial infarction. ACS Biomater Sci Eng 8(8):3271–3298

Terrasini N, Lionetti V (2017) Exosomes in critical illness. Crit Care Med 45(6):1054–1060

Tieu T, Alba M, Elnathan R, Cifuentes-Rius A, Voelcker NH (2019) Advances in porous silicon–based nanomaterials for diagnostic and therapeutic applications. Adv Ther 2(1):1800095

Toffoli G, Hadla M, Corona G, Caligiuri I, Palazzolo S, Semeraro S, Gamini A, Canzonieri V, Rizzolio F (2015) Exosomal doxorubicin reduces the cardiac toxicity of doxorubicin. Nanomedicine 10(19):2963–2971

Tölli MA, Ferreira MP, Kinnunen SM, Rysä J, Mäkilä EM, Szabó Z, Serpi RE, Ohukainen PJ, Välimäki MJ, Correia AM (2014) In vivo biocompatibility of porous silicon biomaterials for drug delivery to the heart. Biomaterials 35(29):8394–8405

Yang J, Wang P, Jiang X, Xu J, Zhang M, Liu F, Lin Y, Tao J, He J, Zhou X, Zhang M (2023) A nanotherapy of octanoic acid ameliorates cardiac arrest/cardiopulmonary resuscitation-induced brain injury via RVG29- and neutrophil membrane-mediated injury relay targeting. ACS Nano

Yao M, Ma M, Zhang H, Zhang Y, Wan G, Shen J, Chen H, Wu R (2018) Mesopore-induced aggregation of cobalt protoporphyrin for photoacoustic imaging and antioxidant protection of stem cells. Adv Funct Mater 28(47):1804497

Zanjanizadeh Ezazi N, Ajdary R, Correia A, Mäkilä E, Salonen J, Kemell M, Hirvonen J, Rojas OJ, Ruskoaho HJ, H. l. A. Santos, (2020) Fabrication and characterization of drug-loaded conductive poly (glycerol sebacate)/nanoparticle-based composite patch for myocardial infarction applications. ACS Appl Mater Interfaces 12(6):6899–6909

Zhang D-X, Yoshikawa C, Welch NG, Pasic P, Thissen H, Voelcker NH (2019) Spatially controlled surface modification of porous silicon for sustained drug delivery applications. Sci Rep 9(1):1367

Chapter 6
Conclusions and Perspectives

Abstract This chapter summarises the applications of nanomedicine in diagnostics and therapeutic approaches for cancer, neurodegenerative diseases, cardiovascular diseases, and orthopaedics related bone tissue engineering.

Keywords Nanomedicine · Cancer · Cardiovascular diseases · Orthopaedics · Neurodegenerative diseases · Drug delivery · Nanomaterials · Combination immunotherapy · Multifunctional nanoparticles · Liposomes · Targeted delivery

In this Brief, we have extensively covered different types of emerging nanomaterials and nanoparticles (NPs) as theranostic tools for cancer, neurodegenerative, orthopaedic, and cardiac diseases. Over the years, there have been enormous research efforts in nanomaterials, aiming to create new scientific fields with interdisciplinary approaches covering chemistry, physics, bioengineering, and medicine. However, the field of nanomedicine is still in its infancy state, which necessitates active collaboration with different fraternities with diverse expertise to tackle the challenges of these burgeoning diseases.

Cancer remains an elusive disease and therefore it is a major health challenge worldwide. Nanomedicine has greatly contributed to cancer in the form of imaging agents, and drug delivery carriers. However, there are many challenges to be addressed. The tumor microenvironment is a key factor which helps the cancer cells proliferate, invade with metastasis and drug resistance. This evades the cancer cell to be killed by the common treatments such as chemotherapy, radiotherapy, and surgery. This necessitates the development of a cancer therapy resistance especially in solid tumors. Recently, cancer nanomedicine offers a great advantage in targeting the tumor microenvironment and treating drug resistance through the advancements in NPs (Sa et al. 2023).

Importantly, NP-based delivery systems have been developed for cancer immunotherapy. Combination immunotherapy is a growing field, which uses the NP-assisted therapies (photothermal, photodynamic and radiotherapy) for targeting and controlling the immunosuppressive cells (including T-cells, dendritic cells, tumour-associated macrophages, etc.) in the tumour microenvironment (Nam et al. 2019;

Yoon et al. 2018). Mesoporous silica NP still emerges as a potential platform for incorporating MEK inhibitor (MEK: mitogen-activated protein kinase enzyme) and anti-PD-1 antibody (PD: programmed cell death protein; checkpoint protein on immune T cells) for combined targeted therapy and immune checkpoint blockade (Liu et al. 2019).

Novel genetic nanomedicines based on a multifunctional nanodevice (lipid NPs, GALA peptide and siRNA) enabled selective lung targeting by the judicious control of lipid NP composition, and improved lung endothelium CD31 gene silencing, thereby achieving efficient therapy for metastatic lung cancer (Abd Elwakil et al. 2019).

The MRI is a powerful non-invasive tool for early and accurate diagnosis of Alzheimer's disease (AD) with behavioural changes and cognitive impairment. Recent years have witnessed significant research activities in the development of NP-based contrast agents functionalized with antibodies, peptides or small molecules for brain MRI (Azria et al. 2017). The ability to target amyloid plaques depends on the size, composition of NPs and their ability to cross the blood–brain-barrier (BBB).

Another emerging nanomaterial is black phosphorus. It can serve not only as a carrier for drugs but also as a neuroprotective agent for neurodegenerative diseases (Ge et al. 2019). Besides brain biomarkers, peripheral blood and skin tissues emerge as potential prognostic biomarkers for probing the disease pathology in neurodegenerative diseases. The application of nanomedicine in the form of scaffolds, drug delivery and imaging systems has effectively improved the neural stem cell-based treatments for neurodegenerative diseases.

As regards orthopaedic and cardiac diseases, nanomedicine-assisted stem cell therapy is an interesting paradigm. One of the notable findings in bone regeneration is the demonstration of articular cartilage implant for osteoarticular regeneration, which utilises nanofibrous poly-ε-caprolactone (PCL) and BMSCs in alginate/hyaluronic acid hydrogel (Keller et al. 2019). This work may enter into phase I clinical trials with the potential of treatment for osteochondral defects, tendon degeneration and age-related musculoskeletal degenerative issues.

Nanomedicine provides advantages in imaging-guided stem cell transplantation and therapy. A notable work is on a multifunctional nanoplatform encompassing mesoporous silica NP and cobalt protoporphyrin (antioxidant drug) for photoacoustic imaging-guided cell delivery and antioxidant protection of stem cells (Yao et al. 2018). This method allows for monitoring specific labelling and delivery of stem cells in myocardial tissues with the advent of photoacoustic imaging.

Current gene editing technology based on engineered nucleases such as clustered regularly interspaced short palindromic repeat (CRISPR)–Cas nucleases combined with Au NPs paves a safer way in treating numerous diseases associated with stem and progenitor cells (Shahbazi et al. 2019) This combined nano/gene editing approach obviates the difficulties associated with the current cellular entries such as electroporation, and virus transduction. Nanotechnology mediated gene editing emerges as an efficient delivery vehicle for CRISPR nucleases of therapeutic interest.

Although nanomaterials provide numerous benefits, we cannot rule out their disadvantages. For instance, a recent study cautioned the negative side of carbon nanotubes

(CNTs). When multiwall CNTs (MWCNTs) introduced into the CNS, they induced higher expression of neuronal nitric oxide synthase in cardiovascular medulla, attenuating sympathetic nerve activity and causing hypotension (low blood pressure and heart rate) (Ma et al. 2018). On the other hand, black phosphorus nanosheets can be used to protect neuronal cell from damage by harvesting Cu^{2+} ions with their strong binding, thereby reducing the cytotoxic ROS generation and copper dyshomeostatis (Ge et al. 2019).

Nanomedicine has attracted widespread interest in recent years, thanks to the emergence of novel nano/biomaterials. Some of the recent advancements include 2D nanomaterials for photothermal therapy (Lin et al. 2017; Cheng et al. 2020; Liu et al. 2020a, b, c), multifunctional Au-based nanomaterials for enhanced cancer radiotherapy (Zhang et al. 2018; Wang et al. 2023), chemo/photothermal synergistic cancer therapy (Liu et al. 2020a, b, c; Chang et al. 2022; Ouyang et al. 2023), and synergistic hyperthermia/immunotherapy (Chang et al. 2021).

A brief introduction (Chap. 1) is given to nanomedicine, an emerging paradigm intersecting two burgeoning fields of nanotechnology and medicine. Different functional nanomaterials, nanocomposites and nanostructures are discussed for diagnostics and therapeutic applications of cancer, cardiovascular, orthopaedics, and neurodegenerative disorders.

In Chap. 2, we have delineated various multifunctional QDs, magnetic NPs and Au-based nanostructures for bioimaging and therapy. Inorganic NPs or organic polymer mediated drug delivery vehicles and liposomes for targeted delivery are discussed. We have also discussed cancer nanomedicine and their promises in preoperative therapeutics, neoadjuvant radiotherapy, chemotherapy, phototherapy, and immunotherapy.

In Chap. 3, we have discussed different nanomedicine approaches for neurodegenerative diseases such as Alzheimer's disease (AD). Several NP-based strategies (e.g., Au-curcumin, PEG-Au NPs,) are addressed to prevent amyloid fibrillation and tau protein hyper-phosphorylation, so that apoptosis and neurodegeneration can be inhibited. Cerium oxide or ceria (CeO_2) NPs can be used as an effective antioxidant for various neurodegenerative diseases including Alzheimer's and Parkinson's diseases, ischemic stroke, and multiple sclerosis.

In Chap. 4, we have discussed the emerging applications of nanomedicine and tissue engineering approaches in orthopaedics. Bone mimicking scaffolds composed of biocompatible and biodegradable polymers such as polylactic acid and polycaprolactone, and inorganic NPs (e.g., reduced graphene oxide, hydroxyapatite/chitosan) have shown great promises in bone tissue engineering by enhancing scaffold fidelity, osteogenic differentiation, and mineralization.

In Chap. 5, we have discussed the applications of nanomedicine in diagnostics and treatment of cardiovascular diseases (CVDs). Notably, antibody-conjugated magnetic NPs have been used to capture and deliver circulating CD63-expressing exosomes to the infarcted heart tissue in rat models of myocardial infarction (Liu et al. 2020a, b, c) This magnetic-guided accumulation of exosomes in infarcted tissue led to reductions in infarct size as well as improved angiogenesis.

Although nanomaterials have great potential for the diagnosis of cancer, cardiac and neurodegenerative diseases, their therapeutic potential is a big question mark owing to their unknown toxicity concerns in the end. Nevertheless, numerous in-vivo studies demonstrated that they are safe and cleared from the body. Compared to inorganic nanomaterials, organic based liposomes, lipids, and biodegradable chitosan NPs and FDA-approved polymers are quite safe to unravel the mysteries of the biological world.

Interestingly, polysaccharide-based drug delivery systems have been utilized for orthopaedic treatments and controlled delivery of anticancer drugs (Hormozi 2023) Furthermore, naturally occurring biocompatible NPs and polymers can be used to deliver anticancer drugs, proteins, peptides, and other genetic materials such as miRNA, siRNA, etc. (Ahmad et al. 2022).

Despite the promise of nanomedicine for the treatment of cardiac diseases, there are also challenges that must be overcome. For example, the potential toxicity of NPs must be carefully evaluated, and the long-term effects of nanomedicine treatments must be thoroughly studied.

References

Abd Elwakil MM, Khalil IA, Elewa YH, Kusumoto K, Sato Y, Shobaki N, Kon Y, Harashima H (2019) Lung-endothelium-targeted nanoparticles based on a pH-sensitive lipid and the GALA peptide enable robust gene silencing and the regression of metastatic lung cancer. Adv Func Mater 29(18):1807677

Ahmad MZ, Rizwanullah M, Ahmad J, Alasmary MY, Akhter MH, Abdel-Wahab BA, Warsi MH, Haque A (2022) Progress in nanomedicine-based drug delivery in designing of chitosan nanoparticles for cancer therapy. Int J Polym Mater Polym Biomater 71(8):602–623

Azria D, Blanquer S, Verdier J-M, Belamie E (2017) Nanoparticles as contrast agents for brain nuclear magnetic resonance imaging in Alzheimer's disease diagnosis. J Mater Chem B 5(35):7216–7237

Chang M, Hou Z, Wang M, Li C, Lin J (2021) Recent advances in hyperthermia therapy-based synergistic immunotherapy. Adv Mater 33(4):2004788

Chang X, Wu Q, Wu Y, Xi X, Cao J, Chu H, Liu Q, Li Y, Wu W, Fang X (2022) Multifunctional Au modified Ti3C2-MXene for photothermal/enzyme dynamic/immune synergistic therapy. Nano Lett 22(20):8321–8330

Cheng L, Wang X, Gong F, Liu T, Liu Z (2020) 2D nanomaterials for cancer theranostic applications. Adv Mater 32(13):1902333

Ge X, Xia Z, Guo S (2019) Recent advances on black phosphorus for biomedicine and biosensing. Adv Func Mater 29(29):1900318

Hormozi B (2023) Plant polysaccharides for orthopedic drug delivery. Plant polysaccharides as pharmaceutical excipients. Elsevier, pp 513–532

Keller L, Pijnenburg L, Idoux-Gillet Y, Bornert F, Benameur L, Tabrizian M, Auvray P, Rosset P, Gonzalo-Daganzo RM, Barrena EG (2019) Preclinical safety study of a combined therapeutic bone wound dressing for osteoarticular regeneration. Nat Commun 10

Lin H, Gao S, Dai C, Chen Y, Shi J (2017) A two-dimensional biodegradable niobium carbide (MXene) for photothermal tumor eradication in NIR-I and NIR-II biowindows. J Am Chem Soc 139(45):16235–16247

Liu X, Feng Y, Xu G, Chen Y, Luo Y, Song J, Bao Y, Yang J, Yu C, Li Y (2019) MAPK-targeted drug delivered by a pH-sensitive MSNP nanocarrier synergizes with PD-1 blockade in melanoma without T-cell suppression. Adv Func Mater 29(12):1806916

Liu S, Chen X, Bao L, Liu T, Yuan P, Yang X, Qiu X, Gooding JJ, Bai Y, Xiao J (2020a) Treatment of infarcted heart tissue via the capture and local delivery of circulating exosomes through antibody-conjugated magnetic nanoparticles. Nat Biomed Eng 4(11):1063–1075

Liu S, Pan X, Liu H (2020b) Two-dimensional nanomaterials for photothermal therapy. Angew Chem 132(15):5943–5953

Liu Y, Shi Q, Zhang Y, Jing J, Pei J (2020c) One-step facile synthesis of Au@ copper–tannic acid coordination core–shell nanostructures as photothermally-enhanced ROS generators for synergistic tumour therapy. New J Chem 44(44):19262–19269

Ma X, Zhong L, Guo H, Wang Y, Gong N, Wang Y, Cai J, Liang XJ (2018) Multiwalled carbon nanotubes induced hypotension by regulating the central nervous system. Adv Func Mater 28(11):1705479

Nam J, Son S, Park KS, Zou W, Shea LD, Moon JJ (2019) Cancer nanomedicine for combination cancer immunotherapy. Nat Rev Mater 1

Ouyang R, Zhang Q, Cao P, Yang Y, Zhao Y, Liu B, Miao Y, Zhou S (2023) Efficient improvement in chemo/photothermal synergistic therapy against lung cancer using Bi@ Au nano-acanthospheres. Colloids Surf B 222:113116

Sa P, Sahoo SK, Dilnawaz F (2023) Responsive role of nanomedicine in the tumor microenvironment and cancer drug resistance. Curr Med Chem

Shahbazi R, Sghia-Hughes G, Reid JL, Kubek S, Haworth KG, Humbert O, Kiem H-P, Adair JE (2019) Targeted homology-directed repair in blood stem and progenitor cells with CRISPR nanoformulations. Nat Mater 1

Wang Z, Ren X, Wang D, Guan L, Li X, Zhao Y, Liu A, He L, Wang T, Zvyagin AV (2023) Novel strategies for tumor radiosensitization mediated by multifunctional gold-based nanomaterials. Biomater Sci

Yao M, Ma M, Zhang H, Zhang Y, Wan G, Shen J, Chen H, Wu R (2018) Mesopore-induced aggregation of cobalt protoporphyrin for photoacoustic imaging and antioxidant protection of stem cells. Adv Func Mater 28(47):1804497

Yoon HY, Selvan ST, Yang Y, Kim MJ, Yi DK, Kwon IC, Kim K (2018) Engineering nanoparticle strategies for effective cancer immunotherapy. Biomaterials 178:597–607

Zhang X, Chen X, Jiang Y-W, Ma N, Xia L-Y, Cheng X, Jia H-R, Liu P, Gu N, Chen Z (2018) Glutathione-depleting gold nanoclusters for enhanced cancer radiotherapy through synergistic external and internal regulations. ACS Appl Mater Interfaces 10(13):10601–10606

Printed in the United States
by Baker & Taylor Publisher Services